# GLOBALIZATION AND EUROPE'S RURAL REGIONS

# Perspectives on Rural Policy and Planning

Series Editors:
**Andrew Gilg**, University of Exeter and University of Gloucestershire, UK
**Henry Buller**, University of Exeter, UK
**Owen Furuseth**, University of North Carolina, USA
**Mark Lapping**, University of South Maine, USA

*Other titles in the series*

**European Integration and Rural Development**
**Actors, Institutions and Power**
Michael Kull
ISBN 978 1 4094 6854 7

**Rural Change in Australia**
**Population, Economy, Environment**
Edited by Rae Dufty-Jones and John Connell
ISBN 978 1 4094 5204 1

**Keeping it in the Family**
**International Perspectives on Succession and Retirement on Family Farms**
Edited by Matt Lobley, John Baker and Ian Whitehead
ISBN 978 1 4094 0995 3

**Rural Revival?**
**Place Marketing, Tree Change and Regional Migration in Australia**
John Connell and Phil McManus
ISBN 978 1 4094 2471 0

**Rural Housing, Exurbanization, and Amenity-Driven Development**
**Contrasting the 'Haves' and the 'Have Nots'**
Edited by David Marcouiller, Mark Lapping and Owen Furuseth
ISBN 978 0 7546 7050 6

# Globalization and Europe's Rural Regions

*Edited by*

JOHN MCDONAGH
*NUI Galway, Ireland*

BIRTE NIENABER
*University of Luxembourg, Luxembourg*

MICHAEL WOODS
*Aberystwyth University, Wales*

Routledge
Taylor & Francis Group
LONDON AND NEW YORK

First published 2015 by Ashgate Publishing

2 Park Square, Milton Park, Abingdon, Oxfordshire OX14 4RN
711 Third Avenue, New York, NY 10017

*Routledge is an imprint of the Taylor & Francis Group, an informa business*

First issued in paperback 2018



**British Library Cataloguing in Publication Data**
A catalogue record for this book is available from the British Library

**The Library of Congress has cataloged the printed edition as follows:**
Globalization and Europe's rural regions / [edited] by John McDonagh, Birte Nienaber and Michael Woods.
pages cm. -- (Perspectives on rural policy and planning)
Includes bibliographical references and index.
ISBN 978-1-4094-2791-9 (hardback : alk. paper)
1. Rural development--Europe. 2. Regional planning--Europe. 3. Globalization--Europe. I. McDonagh, John.
HN380.Z9C6445 2015
307.1'412094--dc23

2014037302

ISBN 978-1-4094-2791-9 (hbk)
ISBN 978-1-138-54651-6 (pbk)

# Contents

# List of Figures

# List of Tables

# List of Contributors

**Joachim Burdack** is an adjunct Professor at the Leibniz Institute for Regional Geography, Leipzig, Germany.

**María Dolores Domínguez García** is a Researcher at the Grupo de Investigación Economía Ecolóxica e Agroecoloxía (GIEEA), Department of Applied Economy, Faculty of Economic and Business Sciences, University of Vigo, Spain.

**Alexandre Dubois** is a Senior Research Fellow, Nordregio, Stockholm, Sweden.

**Maura Farrell** is a Lecturer in Rural Geography in the Discipline of Geography, School of Geography & Archaeology, National University of Ireland, Galway (NUIG), Ireland.

**Xavier Simón Fernández** is an Associate Professor and Chair of Grupo de Investigación Economía Ecolóxica e Agroecoloxía (GIEEA), Department of Applied Economy, Faculty of Economic and Business Sciences, University of Vigo, Spain.

**Violetta Frys** is a Specialist Adviser at the Kreisverwaltung des Saarpfalz-Kreises, Europaangelegenheiten und Partnerschaften, Homburg, Germany.

**Moa Hedström** is a Strategist, Municipality of Skellefteå, Sweden.

**Jesse Heley** is a Lecturer in the Department of Geography and Earth Sciences, Aberystwyth University, Wales.

**Lummina G. Horlings** is an Assistant Professor in the Rural Sociology Group, Wageningen University, the Netherlands.

**Laura Jones** is a Researcher at the Department of Geography and Earth Sciences, Aberystwyth University, Wales.

**Emilija Kairytė** is a Researcher and Director of Institute NeVork, Slovenia.

**Michael Kriszan** is a Researcher at the Leibniz Institute for Regional Geography, Leipzig, Germany.

**Barbara Lampič** is a Lecturer in the Department of Geography, Faculty of Arts, University of Ljubljana, Slovenia.

**Marie Mahon** is a Lecturer in Rural Geography in the Discipline of Geography, School of Geography & Archaeology, National University of Ireland, Galway (NUIG), Ireland.

**John McDonagh** is a Senior Lecturer in Rural Geography in the Discipline of Geography, School of Geography & Archaeology, National University of Ireland, Galway (NUIG), Ireland.

**Irena Mrak** is a Lecturer in the Department of Geography, Faculty of Arts, University of Ljubljana, Slovenia.

**Robert Nadler** is a Researcher at the Leibniz Institute for Regional Geography, Leipzig, Germany.

**Birte Nienaber** is an Associate Professor at the Institute of Geography and Spatial Planning, University of Luxembourg, Luxembourg.

**Irma Potočnik Slavič** is an Assistant Professor in the Department of Geography, Faculty of Arts, University of Ljubljana, Slovenia.

**Dirk Roep** is an Assistant Professor in the Rural Sociology Group, Wageningen University, the Netherlands.

**Paul Swagemakers** is a Researcher at the Grupo de Investigación Economía Ecolóxica e Agroecoloxía (GIEEA), Department of Applied Economy, Faculty of Economic and Business Sciences, University of Vigo, Spain.

**Milada Šťastná** is an Assistant Professor at the Department of Applied and Landscape Ecology, Mendel University, Brno, Czech Republic.

**Suzie Watkin** is a Researcher at the Department of Geography and Earth Sciences, Aberystwyth University, Wales.

**Wiebke Wellbrock** is a Scientific Staff Member of the Department of Rural Sociology, Hohenheim University, Stuttgart, Germany.

**Johannes S.C. Wiskerke** is a Professor of Rural Sociology and Chair of the Rural Sociology Group, Department of Social Sciences, Wageningen University, the Netherlands.

**Michael Woods** is a Professor at the Department of Geography and Earth Sciences, Aberystwyth University, Wales.

Chapter 1

# Globalization and Europe's Rural Regions – Challenge and Opportunity

John McDonagh, Michael Woods and Birte Nienaber

## Introduction

Globalization is the dominant force reshaping rural economies and societies at the start of the twenty-first century. Its reach is pervasive, its pace unrelenting and its impacts and implications present both challenges and opportunities for rural regions in Europe. As the European Commission has acknowledged, 'globalization is seen to touch every walk of life – opening doors, creating opportunities, raising apprehensions', such that 'our response to globalization has moved to the heart of the EU policy agenda' (CEC 2007). Yet globalization is not a monolithic movement, or a singular process. Globalization as a concept encompasses a wide range of different tendencies and developments that sometimes act in concert, but can also work independently, or even in tension with each other, but which collectively have a transformative impact on rural regions. These include the integration of economic systems and the increased mobility of goods and capital across national borders, as well as the concentration of economic influence in transnational corporations with a footloose approach to capital accumulation; but also the increased mobility of people, the intensification of communication networks, the stretching of social relations and the circulation and hybridization of culture and knowledge systems. The phrase 'it's a small world' is more relevant now than ever before, encapsulating the way in which our world has become more integrated, interdependent and connected across a whole series of scales – individual, community, national and international.

Globalization, however, does not mean that everything is now organized at the global scale. It is a direction of travel – movement towards the global – that can include new networks linking adjacent border regions in neighbouring countries as well as transactions with businesses located on the other side of the world. Moreover, this relational perspective shows that globalization is not something imposed on regions from the top down, but rather that globalization works through local places and is reproduced at the local scale (see Massey 2005, Smith 2001, Woods 2007). These conditions, together with the internal inconsistencies of globalization, mean that it does not produce a completely homogenous global society. As Sage (2012) observes, the 'transformations wrought by globalisation are more incomplete, discontinuous and uneven in their consequences and efforts than is often presented' (p. 21). Accordingly, the challenge for rural development policy is not just how to respond to globalization, but how to respond to globalization in a fashion that is appropriate to the regional context.

This book investigates this problem by examining the experiences of ten case study regions across Europe, ranging from the arctic periphery of northern Sweden to peri-urban districts of Germany and the Netherlands. The chapters, which draw primarily on research conducted as part of a European Framework Programme 7 project, 'Developing Europe's Rural Regions in the Era of Globalization' (DERREG), outline some of the most significant impacts of globalization evident in the regions, but also engage with rural development strategies and grassroots initiatives that have sought to respond to these challenges. Through synthesis of these individual regional accounts, the book identifies lessons for our critical understanding of globalization and its impacts which ultimately help advance future rural development policy.

## Defining Globalization and the Global Countryside

As Faulconbridge and Beaverstock (2009) observe, the concept of globalization 'has long been contested' (p. 332), as has the interpretation of its inevitability and desirability. Geographers, economists, sociologists and political scientists propose an array of definitions, each differing in their emphasis. In this research, however, we have followed Steger (2003) in characterizing globalization as 'a multidimensional set of social processes that create, multiply, stretch, and intensify worldwide social interdependencies and exchanges while at the same time fostering in people a growing awareness of deepening connections between the local and the distant' (p. 11). This definition, which resonates with the relational approach of geographers such as Ash Amin and Doreen Massey, emphasizes the complexity and contingency of globalization, but is also neutral in evaluating the outcomes of globalization as advantageous or otherwise. Viewed from different perspectives, and within different contexts, globalization can be perceived as variously 'smiling and promising' and 'less attractive if not threatening' (Sikod 2006, p. 1). Debates between these contrasting positions have pervaded globalization scholarship and political responses to globalization, pitching 'hyperglobalists' against 'sceptics' (Murray 2006).

Thus, whilst some commentators align their discussion around the threat of centralization and standardization and observe how globalization 'has become the focus for a wide range of protests against various features of the world economy' (Masson 2001, p. 1), others have highlighted the opportunities created through technology, communication, liberalization and increased mobility in the newly 'flat world' (Friedman 2005). Assessments of the implications of globalization for rural regions, in particular, tend to polarize between wildly optimistic predictions that time-space compression will enable rural areas to compete on an equal footing with cities in a globalized economy, to doom-laded warnings that rural regions will be the inevitable, helpless victims of globalization as power is concentrated in emergent global cities – if, that is, the plight of rural areas is considered at all.

Indeed, Woods (2007) has observed that 'research into the processes, effects and experiences of globalization continues to exhibit a distinct spatial bias' (p. 488), whilst Hogan (2004) similarly noted, 'a discernable privileging of urban over rural in scholarly accounts of globalization' (p. 22). The limited rural research on globalization that did exist, argued Woods (2007), focused on structures and networks, such as transnational commodity chains, or treated globalization as an unproblematized backdrop for rural change, and in comparison with the wealth of literature on the global city, work on the local, place-based experience of globalization in rural areas was scant, contributing to an incomplete mosaic of understanding.

In response, Woods (2007) posited the notion of the 'global countryside' as a geographical and conceptual counterpoint to the 'global city'. The global countryside is presented as a space that has become increasingly integrated and interconnected through globalization processes (and as such is as much a part of globalization as the global city), but which is also contested and marked by the redistribution of power. In contrast to the assertiveness of the global city discourse, the global countryside is depicted as emergent, fragile and fragmented. It is also intrinsically a more diffuse and dispersed space – Woods follows Massey (2005) in conceiving rural places not as bounded territories, but as unique entanglements of diverse social, economic, political and cultural relations. Globalization, it is argued, proceeds by capturing, substituting, cutting and replacing these relations, changing the ways in which rural places are connected to other places, but also altering the composition of rural places themselves:

> The reconstitution of rural spaces under globalization results from the permeability of rural localities as hybrid assemblages of human and non-human entities, knitted-together intersections of networks and flows that are never wholly fixed or contained at the local scale and whose constant shape-shifting eludes a singular representation of place. Globalization processes introduce into rural localities new networks of global interconnectivity, which become threaded through and entangled with existing local assemblages, sometimes acting in concert and sometimes pulling local actants in conflicting directions. Through these entanglements, intersections and entrapments, the experience of globalization changes rural places, but it never eradicates the local. Rather, the networks, flows and actors introduced by globalization processes fuse and combine with extant local entities to produce new hybrid formations. In this way, places in the emergent global countryside retain their local distinctiveness, but they are also different to *how they were before*. (Woods 2007, pp. 499–500)

This relational approach allows us to move 'beyond the flawed perspectives of essentialist accounts that presented places as fixed and bounded, structuralist accounts that denuded the significance of locality in emphasizing super-structures, and social constructivist approaches that prioritized the discursive to the neglect of the material' (Woods and McDonagh 2011, p. 155). With its rejection of the more easily digestible narratives of homogenization and time-space compression, the relational approach follows a path in favour of more meaningful questions around how 'rural places are remade under globalization, about how rural localities are enrolled into new or refashioned relations with other places, and how extra-local relations are captured, manipulated and exploited by rural actors' (ibid., p. 156). It is these questions that formed the framework for the research reported in this book. As such, it is not our intention to add layers to the debate by engaging in 'semantic jousting about precisely what is meant by globalization' (Dicken 2004, p. 5) and become mired in arguments about the merits of globalization, or the question of whether contemporary globalization is something new. Rather, our focus is on analysis of how globalization is experienced, negotiated, contested and appropriated in rural regions.

Our agenda was further influenced by recognition that globalization is 'not simply about one scale (the global) becoming more important than others but about changes in the relationships between geographical scales' (ibid., p. 9), the objective being to provide insight into the varying ways in which globalization impacts within and across Europe's rural regions and to 'illustrate and examine the complex realities of globalization in rural Europe ... (and) the ways in which actors in rural regions are responding to the challenge of regional development in the context of globalization' (Woods and McDonagh 2011, p. 154). This has involved unpacking the many contradictions that globalization presents – for example, the differentiated spatial pattern of economic activities moving with ease from one country or jurisdiction to another, bringing benefits through investment in one region, but unemployment, depopulation and a decline in services in another – and attempting to understand how and why these produce different outcomes in different regions.

## Differentiated Globalization and Rural Regions

The differentiated impact of globalization on rural regions has been noted both in academic research and in policy development. Writing in a Canadian context, Epp and Whitson (2001) suggest that different rural areas have been repositioned 'to serve two new and very different purposes – playground and dumping ground – as the traditional rural economy declines' (p. xv). Aguayo (2008) in addition, distinguishes between *globalized villages*, 'which have lost control over their own process of global hybridization and have become victims of all kinds of "penetrations" – with all its violent phallic meaning – by which capital, cultural products, and symbols, alter the livelihood and way of life in the village' (p. 542) and *global villages*, which are 'exceptional places that have gained control of their process of global hybridization ... actually advancing themselves as spaces of material, cultural and ideological production for the world' (p. 543). In a similar vein, the European Commission report, *Regions 2020: An Assessment of Future Challenges for EU Regions*, states that rural and remote regions are often more vulnerable to the impact of globalization than urban regions, but that some rural regions have the potential to out-perform metropolitan regions (CEC 2008).

However, a simple winner versus loser scenario understates the complexity of globalization impacts in rural regions. In any one region, multiple diverse globalization processes may have both positive and negative impacts, whilst the effect of individual processes may be experienced differently by different social groups or economic sectors within a region. Understanding the outcomes of globalization in any given region, and identifying appropriate policy and rural development responses, hence requires a more sophisticated approach that takes into account the various intervening factors that shape regional exposure to globalization processes and capacity to respond. Figure 1.1 presents an interpretative model derived from the research presented in this book which seeks to do this, positioning regional outcomes of globalization as the result of the complex interaction of global processes, regional contexts and capacities, catalysts and policies and grassroots initiatives (for further discussion of this model see also Woods 2013a, 2013b).

The model acknowledges that there are myriad ways in which globalization impacts on rural areas. Global processes of change include the liberalization of trade, markets and the flow of capital; the stretching and reconfiguration of global commodity chains, facilitated not only by economic liberalization but also by technological advances; the expansion of transnational corporations, engaged in increasingly complex global operations; the convergence of global consumer fashions and preferences; the liberalization of property markets and growth of international investment in land; the accelerated flow of tourists and migrants around the world, enabled by new and less expensive transport and

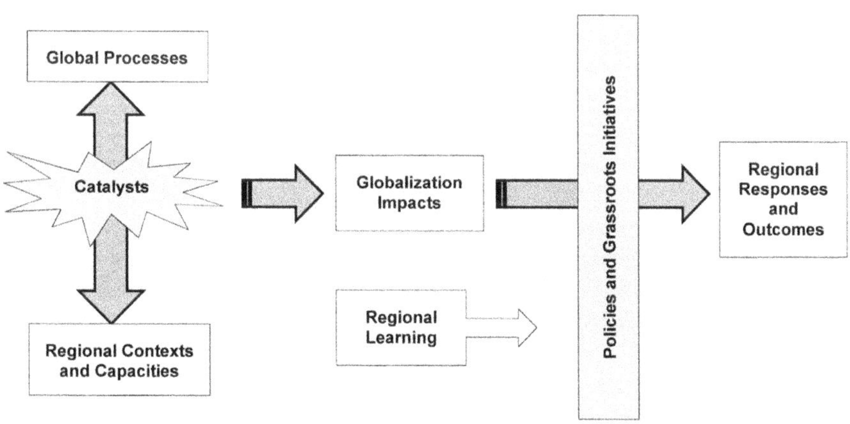


**Figure 1.1 Interpretative model of globalization impacts and regional responses**

communication networks as well as by relaxation of controls on international travel; the internationalization of labour markets in certain economic sectors; the development of new communications technologies, notably the internet, that facilitate instantaneous contact and transactions across the globe; the growth of the global media; the campaigning of transnational pressure groups and NGOs; the adoption of international standards through treaties and conventions; the growth of global consciousness of issues such as climate change, and the promotion of universal values in areas such as environmental management and animal welfare. In theory, any of these processes could impact on any rural region, yet in practice their expression is variegated.

The exposure of a region to any specific global process is filtered by its particular regional context and capacities. Geographical location, for example, still matters – accessibility to international transport links, proximity to national borders (and the relative openness of the border), the size of the regional market, all influence patterns of economic investment and disinvestment and migration, tourism and capital flows. The presence of resource reserves of key minerals, oil or gas is a precondition for enrolment in global energy and resource chains, whilst the significance of regional cultural and heritage resources, its landscape and its environmental amenities shape opportunities for international tourism and amenity migration. Political-economic conditions can also make a difference, including the openness and structure of the regional economy, wage levels and land prices, taxation policies and position relative to major trading blocs such as the European Union, acting variously as both push- and pull-factors for economic investment and labour migration. Finally, the human capital of a region helps to shape opportunities or vulnerabilities, including education levels, the supply of local labour, experience of international travel, diasporic connections and language (especially English).

Taken together, the combination of global processes with regional contexts and capacities determines the potential exposure of a region to globalization, but for actual impacts to occur a catalyst is required. It is the operation of catalysts that helps to explain the local geographies of globalization: why international tourists flock to one mountain village rather than its neighbour; why one town attracts foreign direct investment and not another; why a certain district has more international amenity migrants than the next. Catalysts can take different forms, as examples in the case studies described in this book show. Government policies can be catalysts, such as the role of the Irish government's fiscal policies in attracting international investment in the 1990s and 2000s; but catalysts can also be individuals – corporate executives making decisions about branch plants or product sourcing, or entrepreneurs setting up new export businesses, developing new tourism ventures, or acting as agents for international migrants. Finally, events can also be catalysts – for regions in Central and Eastern Europe, for example, including those in the Czech Republic, Lithuania and Slovenia discussed in Chapters 7, 8 and 11, the fall of communism and later accession to the European Union were catalysts for globalization impacts.

Catalysts make the connections that lead to the types of globalization impacts that are increasingly familiar in rural regions in Europe and beyond, both positive and negative. These might include the opening of new branch plants, or the expansion of local businesses into new markets, but also the closure of factories and markets under pressure of competition from elsewhere; the arrival of international migrant workers, but also out-migration to find work abroad; the development of tourism facilities for international visitors, and the sale of local properties to foreign buyers; the designation of internationally recognized protected areas for conservation, but also challenges from transnational environmental campaigners and agreements to established local economic interests or traditions; among others. However, to finish our analysis at this point would be to reproduce the representation of rural regions as passive recipients of globalization (save for the role of a few local entrepreneurs as catalysts).

Our argument is that local agents in rural regions are active participants in reproducing (or resisting) globalization, because from a relational perspective globalization processes need to enrol local actors and entities to achieve their desired outcomes (Woods 2007). As such, what matters in shaping the outcomes of globalization for rural regions is not just the impact of globalization processes, but also how actors within the region respond to these impacts. These may be individual responses or collective actions; formal regional development programmes and government policies, or the informal mobilization of grassroots social movements; they can include initiatives aimed at harnessing new transnational connections – for example to develop new export markets or to attract international tourists – as well as more defensive strategies attempting to strengthen local economic and social systems and reassert local identity in the face of global pressures. It is these rural responses to globalization that were the primary object for the research reported in this book and which form the focus of the following chapters.

Furthermore, we suggest that regional responses to globalization are a reflexive process, involving internal deliberation about the characteristics of the region, the challenge posed by globalization and the vision for the region's future. Again, such reflexive deliberations can take different forms and be conducted in different settings, but are most effective when conducted in an open and inclusive manner as a focus of collective mobilization and informed by research and evidence – exercises in what have been described as collective learning or regional learning (Adger et al. 2005, Gibney 2011). The development and deployment of regional learning as part of rural responses to globalization is particularly discussed in Chapters 3 and 6 (see also Burdack et al. 2013, Wellbrock et al. 2013a, Wellbrock et al. 2013b).

## The DERREG Project

The chapters in this volume draw on research conducted as part of a three-year project, 'Developing Europe's Rural Regions in the Era of Globalization' (DERREG), which was funded by the European Union's Framework Programme

7 (the exception is Chapter 5, which draws on parallel research undertaken in Wales as part of the Wales Institute for Social and Economic Research, Data and Methods (WISERD) locality programme, funded by the UK Economic and Social Research Council (ESRC) and the Higher Education Funding Council for Wales (HEFCW)). The DERREG research team was led by Aberystwyth University (UK), and included partners from the Leibniz-Institut für Länderkunde (Germany), Mendelova univerzita v Brně (Mendel University, Czech Republic), National University of Ireland, Galway (Ireland), Nordregio (Sweden), Research Institute NeVork (Slovenia), Universität des Saarlandes (Germany), Univerza v Ljubljani (Ljubljana University, Slovenia) and Wageningen Universiteit (Netherlands). More information about the project, including copies of project reports and other outputs, can be found at www.derreg.eu

DERREG aimed to develop understanding of how regional development policies and initiatives can effectively respond to globalization, developing and employing the analytical framework outlined above. In recognition of the diversity of globalization processes impacting on rural regions, the DERREG research was organized around four key themes, each addressing a specific dimension of globalization. The first theme examined the *global engagement and local embeddedness* of rural businesses and, in particular, the transactional and support networks of small and medium-sized enterprises (SMEs). Led by Nordregio, research on this theme involved an e-mail survey of over 200 firms in five case study regions to collect data on the scale and reach of business networks, supplemented by structured interviews with company managers – designed to explore the characteristics and evaluations of interactions in more detail – and 'network brokers' in the public, private and voluntary sectors. The findings from this theme are discussed in Chapters 7, 8 and 11.

The second theme investigated *international mobility and migration to rural regions*, focusing particularly on migrant workers, foreign homeowners and return transnational migrants. Led by the Universität des Saarlandes, the research for this theme involved analysis of statistical and documentary evidence and qualitative interviews with migrants, municipal leaders and representatives of initiatives working with migrant communities. A total of 113 interviews were conducted with migrants across six case study regions, including 50 with migrant workers, 39 with foreign homeowners and 24 with return migrants. The findings from this theme are discussed in Chapters 7, 8, 9 and 10.

The third theme concerned *environmental capital and sustainable rural development*, exploring the incorporation of local and global environmental issues into regional development strategies and opportunities for the exploitation of rural environmental capital through sustainable development. The research for this theme, led by the Leibniz-Institut für Länderkunde, involved the analysis of statistics, policy documents and media reports concerning environmental issues and sustainable development, interviews with regional stakeholders and detailed case studies in each of five study regions. The findings from this theme are discussed in Chapters 4, 8, and 9.

Finally, the fourth theme examined *capacity building, governance and knowledge systems*, particularly by developing and applying the concept of 'rural learning regions'. Research, led by Wageningen University, was conducted in six case study regions and involved the mapping and analysis of policy strategies to support joint learning and innovation – drawing on web searches, literature reviews and expert interviews; the in-depth analysis of a sample of grassroots initiatives in each region, including further interviews; as well as the identification and analysis of 'well-working' examples. The findings from this theme are discussed in Chapters 3, 4, 6, 7 and 10.

Reflecting the emphasis on regional context in the analytical framework, the DERREG research was concentrated in ten case study regions, although not all themes were investigated in each region. The case study regions were selected to provide an illustrative range of geographical settings, from peripheral areas of Spain and Sweden to peri-urban districts in Germany and the Netherlands (see Figure 1.2). They also exhibit differing economic structures and histories, varying levels of GDP and average income, and contrasting patterns of migration and population change (see Table 1.1). These factors in turn mean that eligibility for EU and national regional and rural development programmes varies between the case study regions, with the inclusion of regions in the Czech Republic, Lithuania and Slovenia permitting examination of areas that have only been able to access EU regional development funds since 2004 alongside regions with a history of qualification for successive programmes over a number of years, such as the West of Ireland. The case study regions also vary in territorial size and in population. This to some degree reflects the specific research to be undertaken in the region, and in some examples the case study was initially focused on a larger area before focusing in on a smaller district for detailed empirical research – notably the Eastern Saxony case study focusing in on the Oberlausitz district and the West of Ireland case study focusing in on County Roscommon.

The number of variables included in the selection and description of the case study regions means that they cannot be considered as representative examples of European rural regions, and that it may not be possible to firmly attribute observed differences to specific variables. However, the different case studies do provide illustrative examples of how globalization has differential outcomes in different rural regions and permit examination of the local factors that contribute to this. Each chapter in this volume is focused on one of the case study regions, with Chapter 5 examining Ceredigion in Mid Wales, UK, as an additional comparative example. More details about the individual chapters are outlined below.

**Table 1.1 Comparative statistics for case study regions**

| Case Study Region | Population and Population Density (2007) | Potential Accessibility index (EU mean=100)/ Geographical Context | GDP per head (2011) | Migration trend | Structural Funds eligibility |
|---|---|---|---|---|---|
| Övre Norrland, Sweden | 509,467<br>3.3 person/km$^2$ | 58.1<br>Predominantly rural | €38,000 | Moderate net out-migration | Competitiveness & Employment region 2007–13; Objective 1, 2000–06. |
| Westerkwartier, Netherlands | 59,869<br>173.4 person/km$^2$ | 74.7 (O Groningen)<br>Significantly rural | €64,900<br>(Overig Groningen) | In-migration | Competitiveness & Employment region |
| West Region, Ireland | 415,500<br>30.5 person/km$^2$ | 58.7<br>Predominantly rural | €28,600 | Significant net in-migration | Phasing-in region, 2007–13; Objective 1 region, 2000–06. |
| Comarca de Verín, Spain | 28,598<br>28.4 person/km$^2$ | 43.3 (Ourense)<br>Predominantly rural | €19,300<br>(Ourense) | Net out-migration | Convergence region; Objective 1 region, 2000–06. |
| Alytus County, Lithuania | 177,040<br>32.6 person/km$^2$ | 48.9<br>Predominantly rural | €6,800 | Net out-migration | Convergence region |
| Goriška, Slovenia* | 119,600<br>51.6 person/km$^2$ | 95.4<br>Predominantly rural | €16,300 | Net in-migration | Convergence region |
| Pomurska (Pomurje), Slovenia | 119,537<br>89.4 person/km$^2$ | 73.5<br>Predominantly rural | €11,900 | Small net out-migration | Convergence region |
| Oberlausitz, Germany | 592,000<br>131.6 person/km$^2$ | 79.5<br>Predominantly rural | €23,200<br>(Dresden region) | Significant net out-migration | Convergence region |
| Saarland, Germany (excluding Saarbrücken) | 679,700<br>357.1 person/km$^2$ | 129.6<br>Significantly rural | €31,200 | Small net out-migration | Competitiveness & Employment region |
| South Moravia, Czech Republic | 1,135,000<br>15.7 person/km$^2$ | 74.3<br>Significantly rural | €13,900 | Small net out-migration | Convergence region |
| Ceredigion, UK** | 76,938<br>6.52 person/ km$^2$ | 69.3 (SW Wales)<br>Predominantly rural | €15,900<br>(SW Wales) | Net in-migration | Convergence region; Objective 1 region, 2000–06. |

*Note*: * Region not covered by a specific chapter in this book; ** Not a DERREG case study region. Sources: Eurostat; ESPON.

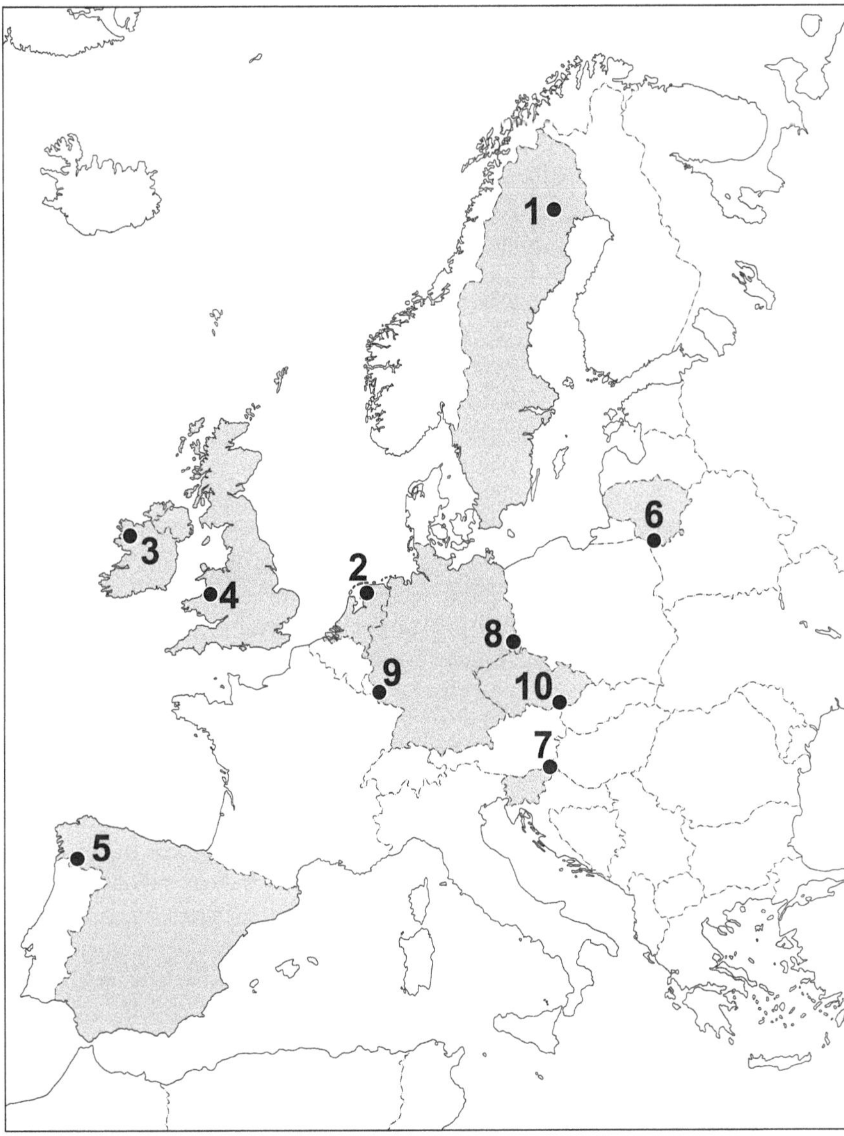


Figure 1.2 Location of case study regions (1: Ovre Norrland, 2: Westerkwartier, 3: West Region, Ireland, 4: Ceredigion, 5: Comarca de Verin, 6: Alytus County, 7: Pomurje, 8: Oberlausitz, 9: Saarland, 10: South Moravia)

## The Structure of this Book

The remainder of this book is structured into ten chapters focused on a specific region and a concluding chapter that presents an overview of the major findings and conclusions from the research. In broad terms, each chapter includes a brief overview of the social, economic and political characteristics of the region and current trends before interrogating the interaction of globalization processes and regional structures and actors, and examining policies and practices that have been adopted to respond to the challenges of globalization. Some chapters adopt a broad perspective, discussing a range of globalization processes that are significant to the region, whilst other chapters focus in on one or two specific issues or initiatives, for example business networking or migration.

The next chapter (Chapter 2), by Dubois and Hedström, explores the significance of the embeddedness and business networks to the success of rural firms in Norrland, northern Sweden, in an effort to produce insights into the scope of local and global relations within rural enterprises and the functioning of networks as innovation-oriented, learning processes of entrepreneurship. Norrland occupies an extreme peripheral location, incorporating a large, sparsely populated interior with the significant urban centres of Umeå, Luleå and Skellefteå located in the coastal strip. The region has a declining population and net out-migration. The traditional forestry industry employs a minority of the labour market, but is significant as an export-orientated sector, with exogenous investment and involvement of extra-regional actors. Particularly notable here is how the double locational disadvantage of sparsity and peripherality is contended with and the impact on the extent to which small businesses in Norrland are integrated in both regional and extra-regional networks.

Chapter 3 focuses on the process of rural regional learning in the Westerkwartier, a district in the northern Netherlands forming part of the province of Groningen. In this chapter, Roep et al. explore the particular expressions of how an area is shaped and reshaped in the context of global forces and how the enhancement of individual and regional learning capacities allows residents of rural regions to become active participants in their negotiation and engagement with global forces while allowing room to develop ideas and strategies that benefit themselves and their region. The Westerkwartier district itself comprises four small rural municipalities close to the city of Groningen, has a growing population and is relatively urbanized in European terms. Agriculture is a significant but pressurized activity and much of the land of the region is classified as protected nature reserve. In shaping the development of the Westerkwartier the authors suggest that systematic and strategic collaboration is played out in the coming together of diverse individuals and groups who may have different perspectives, value sets and in some instances conflicting interests, but who share a willingness to bridge these differences and through informal, visionary leadership have resulted in a collective spirit driving various shared initiatives and a more united representation of the Westerkwartier Region.

Chapter 4, by Mahon et al., focuses on the west region of Ireland. The local context as a force for change as part of wider rural sustainability responses to globalization forms the nucleus of this investigation. Made up of counties Galway, Mayo and Roscommon, this is a classic peripheral location in a European context, and has a relatively sparse population with only a third of the population living in settlements of over 1,500 residents and only one significant urban centre, Galway. Against this backdrop, the chapter examines contemporary civil society in the context of a globalizing rural, specifically the extent to which it can exert agency in determining rural futures under globalization. It focuses particularly on issues of governance, and explores the nature of civil society engagements at various scales and involving different groups and institutional interests. In particular what emerges in this chapter is the need for continued debate on the expectations of civil society in rural development contexts and the necessity to examine the processes of governance, the impact of policy interventions and wider economic agendas, in order to fully appreciate how the local experience of rural sustainability is shaped.

In Chapter 5, Jones et al. use the example of Ceredigion, a rural county within the 'West Wales and the Valleys' convergence region, with a low population density by British standards. The region faces the classic challenges of poor internal and external transportation links, rural isolation and peripherality which present structural factors to be negotiated, alongside multifaceted globalization processes. In using this region as an example the authors suggest that Ceredigion has the potential to provide insights into the ongoing processes of economic restructuring and differentiation occurring across the Welsh countryside due to the declining dominance of agricultural employment and the rise of new modes of economic, social and political relations. Further they suggest that this engagement with globalization processes is also played out in the emphases on local embeddedness and place-specific resources, knowledges and cultural forms, while promoting an outward-looking policy in order to sell this place-specificity to extra-local (national, international) consumers. The fact that large areas of the Ceredigion landscape are afforded protected status due to their exceptional environmental qualities also presents challenges and opportunities where the importance of place is attested to through business practices focused upon selling 'localness' to a global market, and where 'place-branding' is used to tell particular stories about the provenance of food, drink and clothing, as well as experiences in eco- and luxury tourism.

In Chapter 6, Domínguez García et al. delve into how global opportunities and constraints structure traditional landscapes. Using the example of the Comarca de Verín, a small and remote district in Galicia, north-west Spain, the key focus is on assessing the often fragile connections between endogenous knowledge, natural resource management and new business models. The Comarca de Verín consists of eight municipalities situated around the Monterrei Valley in mountains close to the Portuguese border, with a population dispersed in small villages and a strong trend of out-migration, especially of young people. The chapter poses questions about the relation of endogenous development to the global economy, and examines the combining of private entrepreneurship, public support, regional learning and informal networks in the unfolding of a territorial rural development model from the 'bottom up'.

Kairytė, in Chapter 7, examines the experiences of the Alytus Region in southern Lithuania, adjacent to borders with Poland and Belarus. An extensively forested region, Alytus has the lowest GDP per head of the ten case study regions and a decreasing population due to out-migration, including emigration for work outside Lithuania. The chapter details the impact of globalization on socio-economic patterns in the region and discusses initiatives aimed at responding to these challenges and supporting economic development. However, the chapter highlights the structural limitations faced by Alytus, which have restricted its capacity to engage proactively with globalization opportunities.

The globalization of mobility and its effects is a common theme through Chapter 8 (Lampič et al.), Chapter 9 (Burdack et al.) and Chapter 10 (Nienaber and Frys). Here varied insights are given into the importance of the new geographical patterns of flow and activity that have become a feature of globalization. The movement of people and labour as signified here is important in helping us understand the way in which national and international migrants, amenity migrants and trans-border networkers impact on and reconfigure the economic activities and rural enterprises in to which they move or leave. Of significance also in these case study regions is the concept of environmental capital and the potential value that natural resources in some of these regions present in terms of their use as commodities to be developed within broader regional development strategies.

Chapter 8 explores the Pomurje in Slovenia and identifies both similarities and differences in their experience of and engagement with globalization. Pomurje, in the far east of the country bordering Hungary and Austria, has a relatively stable population. Lampič et al.'s exploration of this region suggests that 'geography matters' and spatial location is still particularly relevant despite advanced communications and open borders. They note that globalization impacts are associated overtly with increased cross-border flows (commuting, trading, business networks, migration) but that there remains limited regional resilience at institutional or regional actor level due to structural disadvantages. In Pomurje, transnational networks relate primarily to environmental resources and to amenity migration, but the challenge is to convert these links into economic development.

In Chapter 9, Burdack et al. engage with the regional distinctions and current trends and developments in the Oberlausitz Region in eastern Germany (former GDR). This region forms the eastern part of the federal state of Saxony and is located in a border triangle with the Czech Republic to the south and Poland to the east. The Oberlausitz Region has been affected by three overlapping processes in the last 20 years or so. These overlapping and intertwined processes can be broadly configured as – globalization, Europeanization and post-socialist transition. Undoubtedly these processes have affected demographic and community development, the regional economy and the natural environment to various extents both in terms of opportunities and challenges. What becomes apparent in this chapter is the expansion and acceleration of transnational mobility and the difficulties presented in terms of attracting and retaining a sufficient labour force that has become dependent on foreign migrant mobility in terms of maintaining and enhancing its economy. This is teased out in the chapter from the historical massive out-migration of the workforce which started shortly after German reunification to a reversal and an increasing in-migration of international migrants in more recent times. A second and interrelated aspect of the chapter draws on the strong industrial base of the Oberlausitz economy. This includes examination of open-cast coal mining and the Boxberg brown coal-fired power station as well as insights on the attempts at realignment towards a sustainable use of natural resources, and the promotion of a rural eco-economy.

Nienaber and Frys's contribution in Chapter 10 follows a similar path in that the focus here is also on international migration and the importance of environmental capital as played out in Saarland in western Germany. Saarland is a Federal state and NUTS2 region with strong cross-border links with France and Luxembourg. It is relatively urbanized and includes the city of Saarbrücken, although over one-third of the land area is forest and there are significant rural districts. The region has a restructuring industrial economy, and migration trends have shifted from small net in-migration to small net out-migration. What is evident in this chapter is the many ways in which globalization is played out across trade, investment and mobility, and, in the case of the decline in mining and steel industries, different environmental trajectories fashioned by new regulations informed by globalized values. The significance of foreign capital in terms of the car industry; the fluidity of movement across international borders with France and Luxembourg and the long-term presence of foreign migrant workers especially from Southern Europe and Turkey present opportunities and challenges for this region. Integration is but one such challenge where groups range from those migrants who move in the hope of a better future for themselves and their families to those who wish to invest in properties and are attracted by the more affordable prices of the region. The chapter also addresses issues around environmental capital and in particular the establishment of a UNESCO biosphere reserve (Bliesgau) and the discussion that has emerged around land use restrictions locally and broader connections internationally particularly in the guise of global regulations.

Chapter 11, by Stastná, investigates globalization in the context of South Moravia, a NUTS3 region in the Czech Republic. This region includes part of the White Carpathian Mountains and the Podyjí National Park with almost one-third of its population of 1.13 million inhabitants living in the city of Brno. Agriculture and forestry are significant land uses, but employ only a small proportion of the labour force. It is interesting here to see through this study how the Czech Republic has become more exposed to the forces of globalization across a series of levels including trade, culture and environmental resources, largely due to the opening of its borders in the context of EU membership. This catalyst has encouraged a far more expansive mobility in terms of outward, inward or amenity migrants, as well as creating new transnational economic opportunities. While it is evident in the South Moravian Region that there is somewhat limited networking between local SMEs and the international market place, this must be considered in the context of former socialist central planning and the post-socialist emphasis on FDI. Further, there are also challenges in terms of resources and the part they play in the local rural economy. For example, the small wine-makers of the region are a case in point whereby such small and largely traditional operations are an ideal resource for the local marketplace and provide an ideal opportunity in terms of niche products, yet the challenge is to position such niche production in the global commodity network.

The concluding chapter by Woods et al. draws together the various arguments, challenges and insights as experienced across the case study examples. The diversity of these examples of global engagement; local embeddedness of rural businesses; international mobility and migration of rural populations; environmental capital and sustainable rural development; capacity building; governance, capacity building and knowledge systems in rural regions illustrate clearly the uneven geographies of globalization. Moreover, the findings and implications for future policies and future research on rural areas and on globalization are outlined.

## References

Adger, W.N., Hughes, T.P., Folke, C., Carpenter, S.R. and Rockstroem, J. (2005) Social-ecological resilience to coastal disasters, *Science*, 309(5737), 1036–9.

Aguayo, B.E.C. (2008) Global villages and rural cosmopolitanism: exploring global ruralities, *Globalizations*, 5, 541–54.

Burdack, J., Kriszan, M. and Nadler, R. (2013) Networks of regional learning in rural Eastern Saxony, in Tamásy, C. and Revilla Diez, J. (eds) *Regional Resilience, Economy and Society*, pp. 213–30. Aldershot: Ashgate.

Commission of the European Communities [CEC] (2007) *The European Interest: Succeeding in the age of globalization*. Contribution of the Commission to the October Meeting of Heads of State and Government. Brussels: CEC.

Commission of the European Communities [CEC] (2008) *Regions 2020. An assessment of future challenges for the EU*. Available at: http://ec.europa.eu/regional_policy/sources/docoffic/working/regions2020/pdf/regions2020_en.pdf [accessed: 13 February 2012].

Dicken, P. (2004) Geographers and 'globalization': (yet) another missed boat? *Transactions of the Institute of British Geographers*, 29, 5–26.

Epp, R. and Whitson, D. (eds) (2001) *Writing Off the Rural West: Globalization, governments and the transformation of rural communities*. Edmonton, KY: University of Alberta Press.

Faulconbridge, J.R. and Beaverstock, J.V. (2009) Globalization: interconnected Worlds, in Clifford, N.J., Holloway, S.L., Rice, S.P. and Valentine, G. (eds) *Key Concepts in Geography*, pp. 331–43. London: Sage.

Friedman T. (2005) *The World is Flat: A brief history of the twenty-first century*. New York: Farrar, Straus and Giroux.

Gibney, J. (2011). Knowledge in a 'shared and interdependent word': Implications for a progressive leadership of cities and regions, *European Planning Studies*, 19(4), 613–27.

Hogan J. (2004) Constructing the global in two rural communities in Australia and Japan, *Journal of Sociology*, 40, 21–40.

Massey D. (2005) *For Space*. London: Sage.

Masson, P. (2001) *Globalization: Facts and figures*. IMF Discussion Paper, PDP/01/04. International Monetary Fund.

Murray, W. (2006) *Geographies of Globalization*. London: Routledge.

Sage, C. (2012) *Environment and Food*. London and New York: Routledge.

Sikod, F. (2006) *Globalization and Rural Development in Africa: The case of the Chad-Cameron oil pipeline*, CSGR Working Paper 203/06. Coventry: Warwick University.

Smith, M.P. (2001) *Transnational Urbanism: Locating globalization*. Malden: Blackwell.

Steger, M. (2003) *Globalization: A very short introduction*. Oxford: Oxford University Press.

Wellbrock, W., Nienaber, B., Kriszan, M., Nadler, R., Frys, W. and Roep, D. (2013a). Arranging support for collective learning in rural areas of Germany, in Tamásy, C. and Revilla Diez, J. (eds) *Regional Resilience, Economy and Society*, pp. 195–212. Aldershot: Ashgate.

Wellbrock, W., Roep, D., Mahon, M., Kairytė, E., Nienaber, B., Dominguez Garcia, M.D., Kriszan, M. and Farrell, M. (2013b) Arranging public support to unfold collaborative modes of governance in rural areas, *Journal of Rural Studies*, 32, 420–429.

Woods, M. (2007) Engaging the global countryside: globalization, hybridity and the reconstitution of rural place, *Progress in Human Geography*, 31, 485–507.

Woods, M. and McDonagh, J. (2011) Rural Europe and the World: Globalization and Rural Development, *European Countryside*, 3, 153–63.

Woods, M. (2013a) Rural development, globalization and European regional policy: perspectives from the DERREG project, *Geographia Polonica*, 86, 99–110.

Woods, M. (2013b) Economic entanglements and the re-shaping of place in the global countryside, in Tamásy, C. and Revilla Diez, J. (eds) *Regional Resilience, Economy and Society*, 11–32. Aldershot: Ashgate.

# Chapter 2
# A Spatial Perspective on Small Firm Networking from a Rural Periphery – the Case of Swedish Norrland

Alexandre Dubois and Moa Hedström

## Introduction

The work presented in this chapter investigates the business networks of small firms located in one of the most peripheral, sparsely populated regions of Europe, Norrland in northern Sweden. The conceptual starting point for this study lies in the investigation of how two networking processes, *local embeddedness* and *global engagement*, may co-exist, or even overlap, in the Norrland business community. Recent findings in rural studies (for example Atterton 2007; Fløysand and Sjøholt 2007; Ring et al. 2010) have shed light on the 'significance of business networks and embeddedness to the success of rural firms' (Young 2010b, p. 393). As originally devised by Granovetter (1985), embeddedness implies that the actions of an economic agent are essentially shaped through a set of social relations that the agent is engaged in. *A priori*, the focus on embeddedness as a possible process leading to rural economic development implied a possible resurgence of rural values. Indeed, the concept of embeddedness has seemed to be particularly appropriate to the rural context, emphasizing the connection between a business community and the regional territory in a manner that has been conventionally associated with agriculture and reflected in terms such as 'terroir' (Barham 2003). Moreover, embeddedness builds on 'bonding social capital' (Young 2010a) that have already been observed as a strong feature of business networks in peripheral regions (Young 2010a). This specific feature has shaped the way rural firms do business with a 'predominant reliance upon the local markets [...] due to the rural firms' remoteness from extended markets and limited numbers and density of business networks' (Galloway et al. 2011, p. 255). Used as a generic term characterizing the relative geographical isolation of a place, *remoteness*, in our regional case study, takes the combined forms of *peripherality*, the isolation of the region from main external markets, and *sparsity*, the distance between local communities.

## Beyond Local Embeddedness

Lowe et al. (1995) argue that there are few peripheral regions that are able to generate economic development based on purely endogenous processes. Uzzi (1996) and Boschma (2005) explain this limitation by proposing that a strong level of embedded relations in a region may lead to negative externalities of *over-embeddedness* or *lock-in*. In other words, information and knowledge that are going 'in circles' do not enable entrepreneurs to grasp development opportunities. Furthermore, all economies need inputs of financial capital in order to survive and grow, meaning that an entirely locally focused economy is neither feasible nor desirable (Young 2010b). Finally, according to Atterton (2007), 'maintaining a predominance of strong, embedded relations in a local area can prevent firms from capturing global opportunities', in the form of 'external technological and product innovations' (Fløysand and Sjøholt 2007). Consequently, discussions of embeddedness in rural areas cannot be limited to its local representation (Fløysand and Sjøholt 2007), as rural business networks may not necessarily be only grounded in strong local relations, but may also engage in extra-local networks. Yet, the limited possibilities for engaging in such relations, due to relative isolation and remoteness, can be seen as the main development paradox for rural, peripheral regions.

Several studies have recently undertaken analysis of the spatial dimension of the rural business networks, especially by focusing on the local/extra-local dichotomy (for example Young 2010b; Atterton 2007; Fløysand and Sjøholt 2007). These studies provide new empirical evidence on the spatial dimension of rural business networking. For instance, Fløysand and Sjøholt (2007) show that rural businesses, although strongly embedded in local networks, have nonetheless been able to establish and maintain close relations with other extra-local actors despite unfavourable geographical conditions. Furthermore, Young (2010b) demonstrated that local and extra-local ties can co-exist within a local rural economy, but that they often show little overlap, as it appears that some firms have mainly local networks, and others have extra-local ones. Finally, Atterton (2007) argued that dominating local networking is associated with the risk of over-embeddedness, leading to the further isolation of the local economy from its regional business environment.

As observed by Young (2010a), the spatial dimension of rural firms' business networks, that is, the combination of both local and extra-local relations, is still under-researched. Interestingly enough, when this spatial dimension has been investigated, it has often been in the case of rural communities located in peripheral, sparsely populated regions, for example Western Canada (Young 2010b), Highlands and Islands of Scotland (Atterton 2007) and Western Norway (Fløysand and Sjøholt 2007). As such, rural communities coping with the problematic of double remoteness, that is, combined peripherality and sparsity, have become pertinent study objects in economic geography for investigating the impact of geographical remoteness on the spatial dimension of rural firm networking.

## Changing Forms of Rural Networking

There is an increased understanding that the nexus of rural development and globalization needs to be investigated through a relational approach, that is, by analysing how the new global conditions are shaping the way rural actors are interacting with their surrounding and beyond. In that respect, Atterton (2007) argues that:

> While rural areas lack the density of firms found in many urban business clusters where networking is the norm, they tend to exhibit stronger social or informal networks than urban areas. Networks help to overcome some of the disadvantages of a peripheral location by serving as a means of achieving economies of scale (for example, in accessing distant markets) or as a source of support, information and knowledge (especially given the absence of more formal information sources in many sparsely populated areas). (Atterton 2007, p. 230)

In recent empirical research, a particular emphasis has been put on the analysis of intra-regional business networks and the process of *local embeddedness* in rural economies (Akgün et al. 2010, Atterton 2007, Fløysand and Sjøholt 2007, Ring et al. 2010, Young 2010b). According to Young, 'the notions of embeddedness and business networks are attractive to rural scholars because they establish both that place matters and that rural places possess special assets that allow them to overcome disadvantages of location and size' (2010b, p. 394). Moreover, he advocates that 'embeddedness and business networks have played a substantial role in ongoing debates about the distinctiveness and economic advantages of rural enterprises in global capitalist economy' (ibid.). This is due to the fact that embeddedness is grounded on interpersonal relations and informal networks that 'have traditionally been stronger and more important in rural than urban areas' (Atterton 2007, p. 229). Hence, the notion of embeddedness provides a conceptual framework for understanding rural economic growth in globalization that leads to more favourable prospects for the development of rural regions.

An important issue raised in this body of research is the impact that globalization has on this local embeddedness of rural firms. For Fløysand and Sjøholt (2007), globalization has not, at least, impacted negatively this process:

> Economic geographers also claim that even if globalisation, in the form of 'time–space compression' (Harvey 1989), has disrupted established geographical patterns of economic relations and reduced space-related costs, firms continue to be embedded in particular regions for sociocultural reasons. (Fløysand and Lindkvist 2001, p. 207)

In a study on the use of the internet by small firms in rural Scotland, Galloway et al. (2011) provide more concrete inputs to this discussion. Indeed, they acknowledge that if 'the internet is further theorized as having the capacity to reduce the disadvantages inherent to rural location, such as isolation from markets, relatively less networking and support provision' (p. 255), while also making clear that empirical evidence shows that ICT often promotes greater integration of internal local markets before integration with larger external markets. Consequently, whereas the expected impact of internet use, as a symbol of globalization, on the networking of rural firms would be increased extra-local integration, the observed processes tend to favour the integration of previously fragmented rural business communities to increase the level of *local embeddedness*, through tighter localized ties.

In an essay sketching a framework for understanding the development of rural areas in globalization, Young tempers this idea of a positive correlation between globalization and embeddedness, as the traditional values of rural areas, rather informal, may clash with the promotion by global processes of more formal sets of values and institutions as 'many rural communities and SMEs are disadvantaged by habits and structures established in traditional economies' (2010a, p. 839). Moreover, he claims that 'although rural areas may have high levels of strong-tie or "bonding" social capital, recent research suggests that these types of connections are less useful than outside ties in generating new economic and entrepreneurial opportunities' (ibid. p. 840). Hence, the main issue with respect to rural development and globalization is not the importance of locally embedded relations, rather the necessity but difficulty for rural actors to access international markets.

The need for peripheral small firms to connect with external actors is confronted with the reality of their relative physical isolation. In that respect, a relevant question to address is if geographic peripherality actually engenders a higher degree of relational marginalization. On that account, the recent studies provide pertinent starting points for reflection.

Atterton (2007), in her study of three rural communities of Northern Scotland, shows that rural firms in the most remote community of the three are more focused on local networks and less open to extra-local networks than communities closer to the regional centre (Inverness). In a similar fashion, the 'findings [...] suggest that distance and isolation restrict opportunities for rural SMEs to access undifferentiated extra-local markets, and spur an overreliance on single or few network connections into broader economies' (Young 2010a, p. 844). These findings suggest that peripherality has an impact not on the capacity of certain firms to develop extra-local ties, but on their capacity to sustain them over time and develop simultaneously multiple extra-local ties. If these few targeted extra-local relations enable firms to 'counter the costs of isolation', their resulting business network nonetheless becomes overly vulnerable, as it leaves them as rather marginal and weak actors in global networks (ibid. p. 851).

While not dismissing the importance of locally embedded interactions, the development of extra-local 'weak' ties may have a substantial leverage effect on the local economy (Atterton 2007; Young 2010b). Yet, Young also suggests that these two spaces can hardly be distinguished functionally:

> Firms that depend on an extra-local clientele and revenues nonetheless routinely draw on local resources to gain and maintain access to these markets. These are not disembedded or footloose entities: a great deal of extra-local success appears to be rooted in informal, reciprocal and everyday exchanges at the local level. (Young 2010b, p. 405)

## The Regional Context of Norrland

The Norrland Region consists of the two northernmost counties of Sweden, *Västerbotten* and *Norrbotten*. In European terms, the region is one of the most peripheral and sparsely populated regions of the continent (Gløersen et al. 2006). This combination of *sparsity* and *peripherality* makes the region, from an economic geography perspective, a very challenging business environment. It constrains the possibility for regional economic agents to engage in both tight intra-regional networks and targeted extra-regional ones and thus limits ability to draw from both external and internal economies of agglomeration. On the one hand, sparsity, which relates to the scattered pattern of small settlements found in the rural parts of Norrland, prevents small businesses from fully drawing on the advantages of embedded relations at the regional scale, as the geographical distance between actors limits the possibility to have frequent, face-to-face interactions. On the other hand, peripherality, which relates to the long distance to reach out to the main national and international markets, constrains the access of local economic agents to larger, extra-regional markets and thus their potential engagement in extra-regional networks. Yet, despite these locational disadvantages, the region as a whole has witnessed fair levels of socio-economic performance during the last decade (cf. Lindqvist 2010).

Our hypothesis in this chapter is that the business networks developed by small rural businesses in Norrland can partly explain this relative success. We additionally argue that many other characteristics of the business environment, rather than the individual firms, such as social and human capital, natural resources, historical legacy, regional policy and welfare state are also instrumental to this success.

### *A Strong Industrial Legacy*

The economy of Norrland was traditionally primary sector dominated (mainly mining and forestry, farming activity is marginal). Today the mining and forestry industries are still important, but, like everywhere else, service activities have grown substantially. These are concentrated in the towns and larger villages. However, rationalization in the primary industry (especially mining and forestry) has reduced their need in local employment, whereas their production has, at the same time, increased.

Entrepreneurship and inward investment tend to be concentrated in the larger urban areas, especially Umeå and Luleå, both of which have research and development facilities associated with universities. Recently, the mining areas have

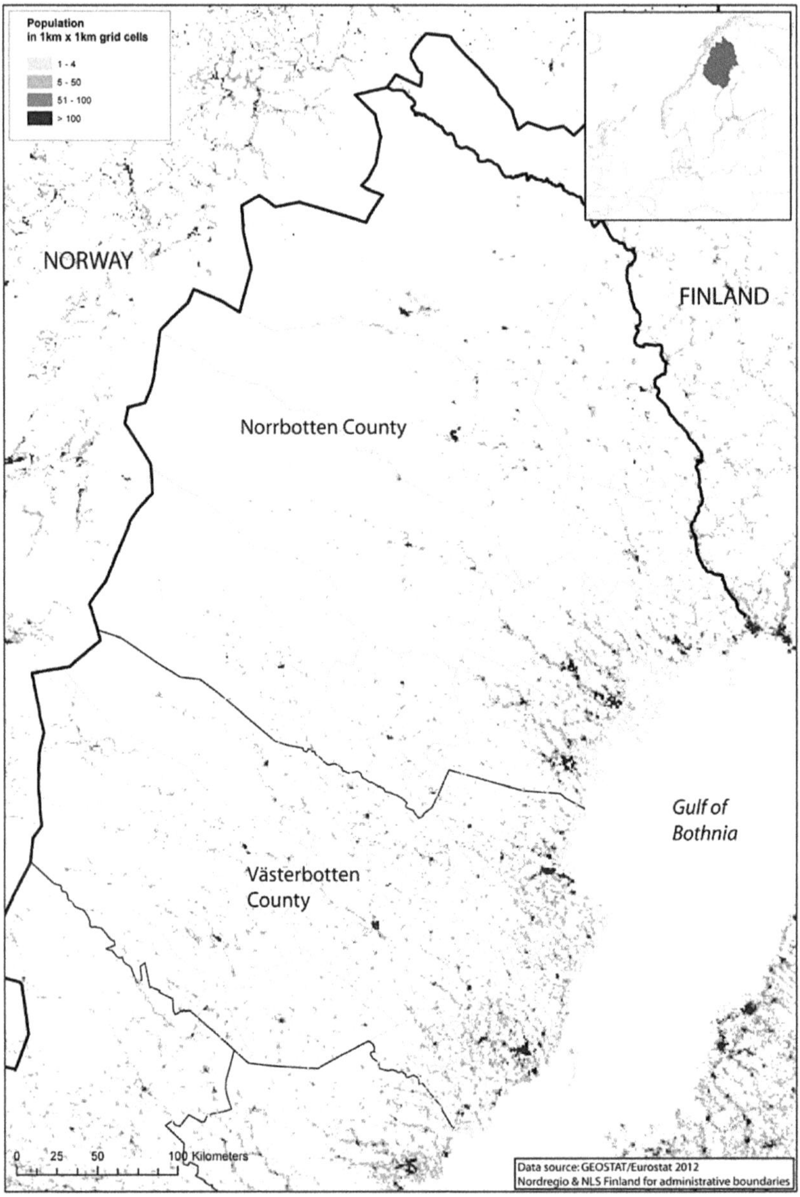


**Figure 2.1 The geographical context of the Norrland case study**

received huge investments since new mines were opened (for example in Pajala). Symptomatic of the importance of mining industries for local development, a large part of two cities (Kiruna and Malmberget) are being moved to new locations nearby, requiring new urban planning, in order to leave space for the further development of the mines (Nilsson 2010). The mines in the region are operated by either large Swedish companies (for example LKAB, Bolliden) or large foreign companies (for example Gold-Ore Resources from Canada and Dragon Mining from Australia).

*A Tradition of Cross-border Integration*

The issue of international engagement of Norrland economic actors may be related to engagement in cross-border ventures with actors in Northern Norway and, to a lesser extent, Northern Finland. This engagement can be attributed to a legacy of cross-border cooperation in the north of the Nordic countries, which dates back to the formalization of Nordic cooperation in the 1950s. As a result, a sense of shared identity emerged and the integration of rules and regulations made interactions between peripheral actors across the border easier. Consequently, a certain degree of shared institutional proximity (Boschma 2005) can be witnessed in this cross-border area.

*Development Challenges and Opportunities in a Global Context*

Understanding the development challenges and opportunities of Norrland in the modern global economy needs to take into consideration both the specificities of the region relating to its geographical location and natural assets, and the particular combination of socio-economic characteristics and policy-industrial legacies. These specificities, we argue, although potentially counteracting one another, have profound impacts over how small businesses engage in regional and extra-regional networks.

As for all rural regions, the development of adapted transport infrastructure has become central in addressing the challenges of double remoteness for Norrland. Overall, one cannot say that the transport infrastructure in Norrland is inadequate, the question being more so whether it is adapted to the specific needs of rural business communities. As an example, the rail transport network is essentially shaped in a north-south axis. The old lines, constructed for, and adapted to, the shipment of the natural resources from the inland mining communities to southern Sweden pass through the inland, parallel to the coast. New investments in a modern railway network, developed along the coast of the Gulf of Bothnia (cf. *Botniabanan*), are more adapted to commuting and passenger transport, and connect the main coastal towns and cities in an attempt at creating a larger regional economy by bringing local economies closer (known as *regional enlargement* or *regionförstoring* in Swedish). The road network provides both inland-coastal and inland-inland connections, but long physical distances between settlements are an obstacle towards a greater integration of rural local labour-markets. Moreover, recent studies have shown that the commuting behaviour in Norrland resembles commuting behaviour in the rest of Europe, with a willingness-to-travel for 45 minutes (Gloersen 2009, Sandow 2008), thus making it unlikely that commuting between rural settlements can be achieved on a daily basis. Consequently, for rural business communities, these 'classic' infrastructure investments may not provide a substantial leverage effect for local economies.

Swedish regional policy has traditionally been strongly focused on the development of its most sparsely populated regions by, for example, having a dedicated development agency for these territories (*Glesbygdsverket*). Yet, the merging of this agency in 2009 into the Swedish Agency for Economic and Regional Growth (*Tillväxtverket*, and formerly *NUTEK*) is symptomatic of the shift in Swedish, and European, regional policy towards growth-oriented regional development initiatives. Since the accession of Sweden to the EU in 1995, the upscaling in financial instruments available have enabled regional policy initiatives to target more progressive approaches to open up rural local economies, by taking advantage of the global improvements in communication networks following two main tracks. The first one concerns the development of air transportation services in a dozen airports throughout the region. These services provide connections from local airports to the national hub (Stockholm), and some intra-regional connections between the main regional centres (Umeå-Luleå-Kiruna). If the star-shaped structure of the air transport network reflects a core-periphery structure, thus increasing their dependence on the capital region (Gløersen et al. 2006), the hub-system provides increased access to more distant destinations. This enables entrepreneurs to source important resources, not least in terms of capital and investments, and reach out beyond the domestic market. The second main initiative to open up local economies of Norrland has been the early and massive investments in the development of ICT infrastructure (mobile telephony, broadband internet) since the 1990s. A consequence of such initiatives is that Sweden is currently the European country with the smallest digital divide between its densely populated and thinly populated regions (European Commission 2010). Penetration levels reach 80 per cent of households, despite the fact that the thinly populated areas are much more widespread than any EU country (with perhaps the exception of Finland). Consequently, public policies in Norrland play an important role in providing local entrepreneurs with the possibility of taking advantage of the technological progress arising from globalization.

Norrland has a business structure which is very dependent upon resource-based activities, such as mining or forestry. Historically, this has been an important factor for the development of the welfare of the region, but also of the Swedish society at large, as the income generated by such industries have fed into the welfare system through taxes. In that respect, Norrland belongs to a category of rural areas that is described by Young (2010a) as 'heavily integrated into global economies because of their role as commodity-producing regions' (p. 838). However, this may also be seen as a structural weakness for the region. First of all, the wider regional economy tends to be vulnerable to fluctuations in demand, and long-term growth prospects may be limited. Second, the presence of large international commodities may end up imposing 'their' networks upon smaller local economic actors and potentially locking them out of the extra-regional networks they actually need (Marsden et al. 2000, Young 2010a).

Finally, as we argue, the capacity of rural businesses to engage in such extra-local connections is also strongly shaped by the regional and national institutional legacies. Young (2010a) suggested that what may be holding back rural businesses from 'going extra-local' are 'a distinct lack of experience', and a closed attitude to external networking. In that respect, the openness of the Swedish economy as a whole can be seen as an advantage for Norrland compared with other rural regions. Indeed, the small size of the domestic market and the abundance of natural assets have led the Swedish economy to be open and export-oriented. Consequently, Swedish economic actors are no stranger to doing business with actors located beyond the national market. Moreover, regional policy initiatives have, in the last decade at least, strongly supported entrepreneurship and innovation in all parts of the country, and provided an active policy supporting the development of SMEs, which constitute the main bulk of businesses, especially in rural areas (Swedish Government Office 2007). Ultimately, the engagement of Swedish national authorities in supporting businesses has materialized, beyond the traditionally available subsidies, through the creation of a dense network of organizations at national, regional and local levels, providing organizational support to small firms in their networking activities (Dubois and Hedström 2011).

## Empirical Findings

The present case study provides new empirical evidence on the small firm networking behaviour in sparsely populated areas of northern Sweden, and especially through which processes these firms connect to extra-regional actors. The material analysed was collected as part of the DERREG project. The work was divided into three consecutive stages: an electronic survey was completed by 50 managers of rural SMEs; face-to-face interviews carried out with a selection of 15 of these firms; and a dozen interviews were made with local, regional or national organizations involved in business support initiatives in Norrland. The collected data from the electronic survey was compiled in a joint table and associations were analysed using a network analysis package (UCINET). The information compiled from the interviews was transcripted, read and analysed by the authors. The targeted businesses were those with less than 50 employees, located outside the urban municipalities of Norrland and that produce a production output potentially tradable extra-locally. While the survey provided *perception data* emphasizing the relative importance of different types of actors (other small firms, multinational companies, public agencies) originating from different economic environments (regional, national, European) in the firms' network, the interviews with SME managers and semi-public organizations provided a deeper understanding of the process-oriented aspects of networking.

The empirical findings are presented using two parallel narratives highlighting two complementary aspects of business networks: on the one hand, networking is portrayed through transactions and market-driven linkages (*logic of trade*); on the other, networking is portrayed through the lens of a collaborative process based on informal exchanges of information grounded in social and kinship relations (*logic of collaboration*), echoing what Storper (1997) identified as 'untraded interdependencies'. The purpose of making this distinction is not to confront the two logics, but rather to emphasize the fact, *in fine*, these two logics are functionally intertwined in small firm networking, and that this functional integration brings together the regional and extra-regional components of the networks.

## Networking through a Logic of Trade

### *In Search of Market Intelligence*

An important observation from our case study concerns the actual motivations and foreseen benefits for rural small firms to engage in networking activities. During the face-to-face interviews, the SME managers were asked to assess the relative importance of different types of actors according to various network functions (Oerlemans and Meeus 2005). These network functions comprise foreseen benefits linked to competition, market position, product development or capital and investments. Two-thirds of the managers interviewed valued their networks principally for the support they provided to marketing (*Improving market position*), or a combination of marketing and product development (*Improving production process*). This finding is consistent with the identification by managers of customers and suppliers as the most important partners in their network. Indeed, in all 15 interviews made, at least one category of customer and supplier (for example regional customer or European supplier) was mentioned by the manager, and usually scored as very influential in the overall mapping of their business network.

On the importance of customers and suppliers in the overall business network, a manager stated:

> Clients represent an enormous knowledge database about the most advanced research fields, and a good contact with them makes that clients can help out and answer questions and come with valuable inputs when we need it. (Respondent 35)

Hence, the privileged place of customers and suppliers in the firms' business network is due to their good knowledge of market and product issues and also in their capacity to generate new trade relations or consolidate existing ones through their endorsement.

### *An Open Rural Economy*

Being asked to assess the geographical distribution of their customer and supplier relations, it became clear that a large majority of the sampled small firms have established trade relations outside the regional market. Using a methodology adapted from the SME Index of Globalization, developed by the OECD (1997), and previously used in the context of rural peripheries (Herdzina et al. 2004), the index applied to our sample enabled us to categorize SMEs with regard to the geographical distribution of their trade (see Table 2.1).

**Table 2.1 Assessing the degree of internationalization using survey data on market penetration (adapted from Herdzina et al., 2004)**

| Degree of internationalization | % of transactions according to survey data | | | |
|---|---|---|---|---|
| | Regional | National | European | World |
| Low | >80% | | | |
| Medium Low | >50% | <50% | | |
| Medium | | >50% | <10% | |
| Medium High | | >30% | >10%, but <30% | <10% |
| High | | | >30% | >10% (or international >40%) |

**Table 2.2 Index of globalization applied to the small firms from the case study (numeric value refers to the number of firms)**

| Index of globalization | Criteria |
|---|---|
| Regional firms (N=6) | Degree of internationalization of BOTH inputs AND outputs from low to medium low |
| Domestic firms (N=12) | Degree of internationalization of inputs from low to medium AND degree of internationalization of outputs as medium;<br>OR<br>Degree of internationalization of outputs from low to medium AND degree of internationalization of inputs as medium |
| Internationalized firms (N=19) | Degree of internationalization of inputs from low to medium AND degree of internationalization of outputs from medium high to high;<br>OR<br>Degree of internationalization of outputs from low to medium AND degree of internationalization of inputs from medium high to high |
| Strongly internationalized firms (N=10) | Degree of internationalization of BOTH inputs AND outputs from medium high to high (N=10) |

For each firm, the two resulting scales (that is, a scale for firm inputs and a scale for firm outputs) are crossed to produce a typology of small firms (see Table 2.2). The following four categories can be identified:

- *Strongly internationalized firms*: firms with predominantly international customers and suppliers;
- *Internationalized firms*: firms with an important share of either input or output networks sourced outside of Sweden;
- *Domestic firms*: firms with their sales or purchases predominantly sourced in Sweden but outside of Upper Norrland;
- *Regional firms*: firms with most of their sales and purchases with partners in the Upper Norrland Region.

Table 2.2 shows that more than 85 per cent (N=41) of the sampled small firms have developed extra-regional transactional linkages. More importantly, 60 per cent (N=29) have developed *extensive* internationalized transactional linkages, while 40 per cent (N=18) have mainly domestic networks. Hence, establishing relations with customers or suppliers located in other regions is a relatively normal behaviour for remote small firms. This is consistent with the managers' own view on the limitations of the domestic market and the need to export:

> The Swedish market is limited and considering all the raw material in the form of the large forests with exceptionally good quality that we have here, exporting has sort of been incorporated in the concept for a long time now. (Respondent 01)

*The 'Strength' of International Partners*

Face-to-face interviews corroborated the high importance given to international partners in the firms' business network. Indeed, managers generally rated international actors as more important than domestic ones. From the case study, it is possible to deduce that although the interviewees identified about one-third of the actors in their networks as either located in Europe or further afield, the overall 'weight' or value assigned to these international contacts was substantially greater than that associated with the other two-thirds of (regional or national) actors. This finding was supported by a number of statements by the interviewees.

An interviewed manager claimed that:

> Doing business with international actors is very time consuming but it is also exciting and makes us more interesting among actors back home. (Respondent 17)

Moreover, the network of international partners is central for the small businesses in order to identify the upcoming market developments. An interviewed manager on the subject:

> The well-developed personal relations with customers, colleagues, competitors and suppliers all around Europe makes us able to understand and predict the development of the European market. (Respondent 37)

Not only do extra-regional, and especially foreign, ties constitute the main bulk of trade relations, but they are usually perceived by managers as being more valuable and as having a higher leverage effect on the firms' operations. Hence, in many instances, existing relations with international trade partners tend to act as a door-opener to further markets for small firms and provides the necessary market intelligence that enables small firms to anticipate future market developments.

## Networking through a Logic of Collaboration

### *Tight Interaction with Domestic SMEs*

In the survey, respondents were asked to assess, using a Likert-like scale, the intensity of their collaboration with different categories of firms, according to their size (SMEs, larger firms, multinational companies) and geographical location (regional, national, EU/EEA, World). The survey made it explicit that such relations focus on information and knowledge exchange. The compilation of data demonstrated that the main collaborative partners for the sampled firms were domestic, including both national and regional, small firms. This seems to be independent from the geographical orientation of the firms' trade network, that is, domestic or internationalized. For both groupings (see previous section), the national and regional small firms are the most central collaborative partners, thus implying that the spatial dimension of trade and collaborative networks are neither strictly following each other, that is, regional (resp. international) trade engendering regional (resp. international) collaboration, as suggested by Tödtling and Kaufman (1999; 2000), nor totally complementing each other, that is, regional (resp. international) trade engendering international (resp. regional) collaboration, as suggested by Young (2010b).

Hence, there is a strong domestic inertia regarding the collaborative network of small firms. This confirms, to some extent, the *homophily principle* advanced by Glückler (2007) which presumes that firms will tend to network with firms that are similar to them. The shared institutional environment provided by the national and regional level provides a high degree of institutional proximity among small firms, which appears to be more important than geographical proximity alone (cf. Boschma 2005). Indeed, collaborative relations tend to focus on issues that are perceived as shared and of mutual interest between small firms, especially those related to a specific domestic *savoir-faire* in a particular sector, or specific challenges inherent to peripherality. The close collaboration with regional small firms may be a more localized version of this process, thus adding the shared locational conditions as cement for collaboration. Interviews tended to corroborate the importance of trust within the rural business community, and the search by the managers to develop reciprocal, informal collaborative links (thus with no side dominating the other) that builds upon rural social capital. A respondent, for instance, stated that:

> Sometimes the most giving meeting among us in the network are the lunches or the times we meet to watch an ice hockey game. Then we can really talk and give each other good advice. These meetings can also end with new orders and hence new jobs for us. (Respondent 10)

In a similar fashion, another respondent talked about the 'good chemistry' within the local business community, which meant that he:

> can always just cross the street and go and ask the others for help and ideas ... (Respondent 31)

A further manager stressed the importance of nurturing the more informal contacts with small and large firms within the same business sector at regional level. These contacts were said to have grown stronger over time and are built on meetings, phone calls and also e-mails:

> The contacts with these colleagues, and at the same time competitors, are based on trust and have taken years to establish. It needs to be taken care of you cannot suddenly turn your back on them. (Respondent 40)

Finally, a manager expressed the differing roles of regional actors in the overall business network:

> The local/regional network plays another role than the international/national network of clients and suppliers. Those who are close are more like friends, I can phone them for support, advice or ideas. (Respondent 23)

### *The Internationalization of Rural Collaborative Networks*

Whereas small firms with an internationalized trade network still have privilege relations with domestic small firms, our findings show that relations with firms, both large and small, located in Europe are more central in the firms' collaborative network than for firms with a more domestic market focus. This supports the idea that the geographical expansions of trade network and collaborative networks are, to some extent, correlated, and that the establishment of trade ties, for peripheral small businesses, need to be embedded into a more collaborative and social context.

*Central Role of the Domestic 'Support Network'*

An important assumption for our work in the case study was that business networks are not constituted solely of private actors (that is, firms), but also involve relations with non-private actors, such as public agencies, research institutions or trade/sectoral organizations, playing an important role as 'support' organizations.

Based on the assessment by SME managers of the intensity of their relations with support organizations by type (trade organizations, public agencies) and geographical origin (regional, national), two interesting observations could be made. First, all domestic support actors got rated higher than international ones. Although domestic support organizations are strongly influenced and embedded in European regional policy initiatives and programmes, small firms usually only directly interact with domestic actors, which thus epitomize the *perceived* institutional support. Second, the importance of support organizations is rather a matter of *who* rather than *where*. Hence, both national and regional trade organizations are deemed the most more central actors in the wider support network, followed by research and education centres, public authorities and agencies and, finally, business consultants. Hence, geographical proximity does not appear to be the prime factor for seeking institutional support.

*Support Organizations as Network Intermediaries*

What, in reality, seems to play a greater role is the capacity of the support organization to act as intermediaries for facilitating the development of inter-firm networks? The pursuit of fair competition through the European Single Market has made the possibilities for direct subsidies to small businesses more controlled and limited, thus shifting the praxis of support organizations from supporting product and production development to supporting networking activities. This shift is very palpable when investigating the Swedish business support context. Combined with processes of increasing international openness, it appears that, although the internationalization of the firms' network is not, according to the network brokers, the primary option when developing brokered networks, it has become, *de facto*, a dominant part of their work, due to the limited possibilities of networking offered within the region. Indeed, support organizations are instrumental in helping firms to find collaborative partners or new market entries in more distant regions.

On the role that support organizations play for the internationalization of regional firms, a network broker claims:

> It boils down to helping them reaching out. Naturally, the contacts with new clients need to be steered by the firms themselves, but by acting as door-opener we can help them with that. (Network broker in regional business organization)

In that framework, two main kinds of brokering activities can be emphasized: match-making and forum facilitation. Match-making activities usually consist, in a particular firm, of identifying a specific need in terms or market or product development that is not available in their (known) immediate business surroundings. The firm thus initiates contact with a network broker, usually the local business development organization, which will support it in finding the most suitable business partners, whether it is located in the region or abroad. Forum facilitation, on the other hand, consists in a network broker facilitating and providing operational support to the process of collaboration initiated by a grouping of regional firms. In many instances, the aim of such fora is to increase the international visibility and competitiveness of the regional firms on specific, sectoral markets.

## Concluding Discussion

The use of narratives enabled us to put into perspective a process-oriented approach to small firm networking, according to two different logics of interaction. A particular focus of this case study was also to highlight the role of a range of actors, originating from different geographical contexts, and participation in the process shaping these networks. The findings elaborated in the previous section are synthesized in the adjacent illustration (Figure 2.2).

The changing forms of rural networking epitomize the shifting perception of what *distance* entails. Young (2006, p. 254) considers that 'distance ought not be considered merely as the geographic tract that separates locales, but rather as an active combination of natural, technological, and social elements'. This understanding of distance is somehow implicit in our conceptual framework: rural actors are geographically remote, but they may at the same time be close in social terms. The importance given to embedded relations in rural and peripheral settings also feeds into this line of thought: proximity between actors is based on strong local cohesiveness as a response to a geographical context engendering relative isolation. In a way, the importance of globalization relates to the capacity of modern technology to enable the recombination of the physical and social aspects of distance. In that framework, distance, and globalization, is not something that is endured by rural actors, but rather something that they can manage and coordinate in order to create new modes of proximity (Young 2010a) beyond the local territorial context.

Two proximity dynamics seem to be especially important in the case of small firm networking: *social proximity* and *institutional proximity*. Social proximity is closely related to the notions of embeddedness and social capital, and can be defined 'in terms of socially embedded relations between agents at the micro-level. Relations between actors are socially embedded when they involve trust based on friendship, kinship and experience' (Boschma 2005, p. 66). Clearly, social proximity is very close to what was portrayed in the literature as the rural comparative advantage, grounded on informal institutions in interpersonal interactions (Ward and Hite 1998). Institutional proximity widens this understanding by emphasizing the importance of societal norms and values over interpersonal relations. Indeed, 'whereas social proximity has been defined in terms of socially embedded relations

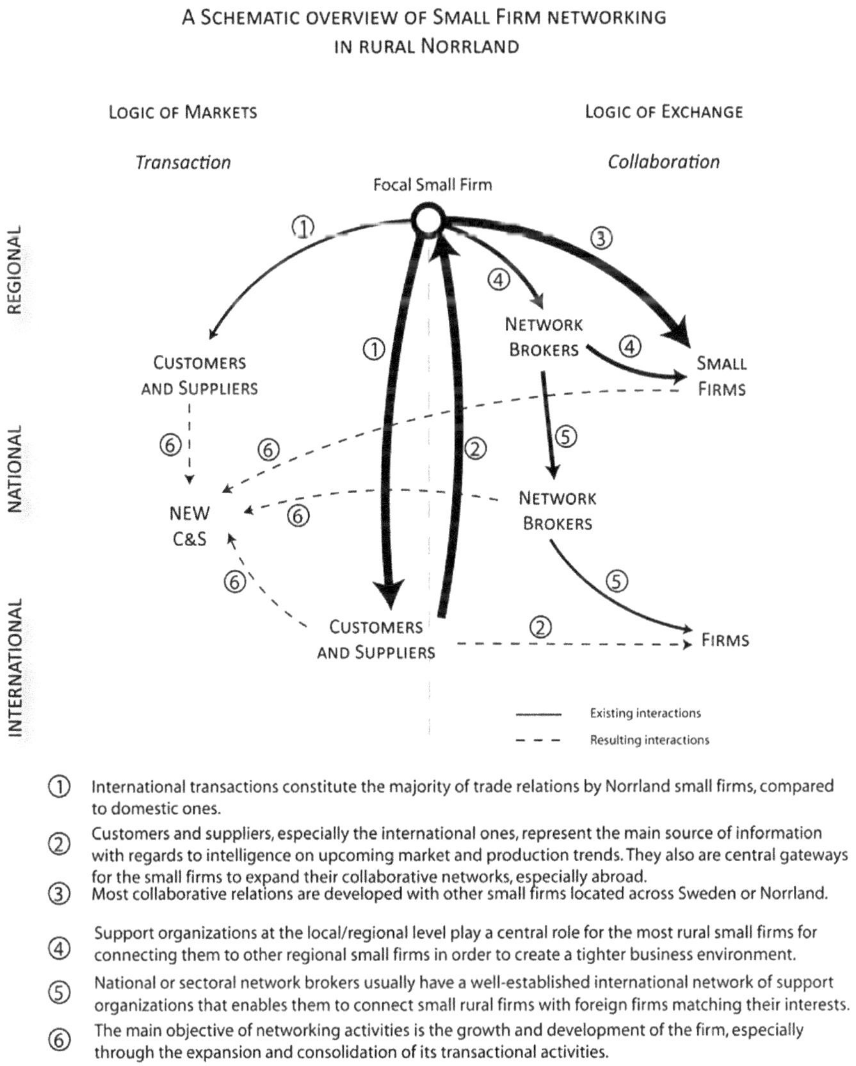


**Figure 2.2** **The two relational logics presented in the narratives highlight a multi-scalar and multi-actor perspective on small firm networking in rural Norrland**

*Source*: authors

between agents at the micro-level (based on friendship, kinship and past experience), institutional proximity will be associated with the institutional framework at the macro-level' (Boschma 2005, p. 67). Boschma claims further that 'the notion of institutional proximity includes both the idea of economic actors sharing the same institutional rules of the game, as well as a set of cultural habits and values' (ibid., p. 68). Our findings seem to show that different proximity dynamics provide different advantages to small firms when interacting with actors at different geographical scales: the regional for social proximity, and the national for institutional proximity.

It was introduced earlier that studies investigating embedded relations in rural settings suggest that local embeddedness is a response by rural actors to the challenges brought about by geographical remoteness. This line of thought however is built on an image of rural communities as isolated 'villages'. In our case study, we argue that sparsity engenders a more diffuse understanding of locality. In concrete terms, it means that even in the local surroundings it takes some effort (that is, costs) for actors to interact with other business partners. As a consequence, 'local' embeddedness alone would not provide sufficient economic return in order to 'overcome the disadvantages of location and size' (Atterton 2007). Our findings highlight the fact that, even if entrepreneurs have a high level of kinship and trust with their local counterparts, they need, in addition, to develop collaborative ties with actors outside the local community.

Our findings also show that geographical remoteness does not hamper the capacity of firms to establish extra-local ties. Indeed, a large majority of the sampled firms have customers and suppliers that are mainly located outside the region. To our understanding, the general open attitude of Swedish businesses and the legacy of investments made in modern communication systems, both transport (regional airports) and Information and Communication Technologies (ICT) (broadband, mobile telephony), have been instrumental in supporting the ability of Norrland entrepreneurs to reach out to extra-regional markets. Moreover, the impact of ICT is also important for strengthening regional networks (Galloway et al. 2011). To a certain extent, ICT has improved the ability of organizing such networks, not least by enabling continued interactions without having to meet face to face. As an example, the large majority of brokered network initiatives have a website acting both as a means to diffuse information outside the (regional) network and as a platform enabling a more efficient interaction process.

Geographical distance is a notable challenge when it comes to intra-regional interactions. Indeed, one can argue that there is little leverage that modern logistics can have for improving the travel time between settlements in the region. For rural actors of Norrland, it is tacitly understood that face-to-face interactions within the region requires more time and engenders higher cost due to travelling than in other (rural) places. What then becomes a disadvantage for Norrland rural actors is that they can hardly benefit from what Young termed 'small-scale "happy accidents"' (2010a, p. 844), and the fact that 'many profoundly important connections emerge by chance or serendipitously in the course of doing "other business"' (ibid., p. 851).

Finally, this research enabled us to understand the important role played by network intermediaries for the vitality of the regional networks and the emergence of new extra-regional 'weak' ties. From our empirical research, it appeared that there are many network initiatives involving small firms of Norrland and brokered by public organizations. As the importance given to customer/supplier relations proves, such initiatives represent, quantitatively, a small part of a firm's business networks. Nonetheless, we argue that such brokered networks enable firms to evolve 'off the beaten track' and engage in relations that they could not have entered by using their existing network partners (for example customers and suppliers) as gateways for creating new contacts. Consequently, we argue that brokered initiatives are important for creating the 'weak' ties that prove to be crucial for the development of small firms. This argument corroborates the conclusions of Atterton following her Scottish case study that '(t)here is a role for policymakers to play (…) as brokers to encourage the formation of extra-local and weak ties' (2007, p. 241).

In conclusion, it seems appropriate to end with a discussion on policies and their potential impact in preparing the regional milieu to participate in globalization processes. The Norrland case study clearly demonstrates the impact that policies may have on the capacity of a rural economy to stay competitive in times of globalization. Although much more needs to be done, proactive regional development policies have given rural businesses essential tools to allow them be connected to global processes, for instance through targeted infrastructure developments. Clearly a reason for the relative success of Norrland stems from the institutional framework which supports not only the development of stronger relations at the regional and national level, but also the establishment of international ties. The Norrland case study demonstrates however that *local embeddedness* may not be sufficient for the development of rural communities. We argue that promoting networking amongst indigenous businesses (cf. Rosenfeld 2002) is, for policymakers, a necessary, but not sufficient, step for fostering economic development in peripheral and rural areas.

## Acknowledgement

The authors would like to thank Andrew Copus (Nordregio) and Brita Hermelin (Stockholm University) for their valuable and insightful comments on earlier versions of this chapter.

## References

Akgün, A.A., Nijkamp, P., Baycan and T. Brons, M. (2010) Embeddedness of entrepreneurs in rural areas: a comparative rough set data analysis, *Tijdschrift voor Economische en Sociale Geografie*, 101(5), 538–53.

Atterton, J. (2007) The 'Strength of weak ties': Social networking by business owners in the Highlands and Islands of Scotland, *Sociologia Ruralis*, 47(3), 228–45.

Barham, E. (2003) Translating terroir: the global challenge of French AOC labelling, *Journal of Rural Studies*, 13, 127–38.

Boschma, R.A. (2005) Proximity and Innovation: a critical assessment, *Regional Studies*, 39(1), 61–74.

Dubois, A. and Hedström, M. (2011) *Striving for Local Development – Brokering firm networks in the North of Europe*, Paper presented at the Regional Studies Association Annual International Conference, 17th–20th April 2011, Newcastle upon Tyne, UK.

European Commission (2010) *Fifth report on economic, social and territorial cohesion.* Consulted November 2011. Available at: http://ec.europa.eu/regional_policy/sources/docoffic/official/reports/cohesion5/index_en.cfm

Fløysand, A. and Sjøholt, P. (2007) Rural development and embeddedness: the importance of human relations for industrial restructuring in rural areas, *Sociologia Ruralis*, 47(3), 205–27.

Galloway, L., Sanders, J. and Deakins, D. (2011) Rural small firms' use of the internet: From global to local, *Journal of Rural Studies*, 27(3), 254–62.

Gloersen, E. (2009) *La centralité à la marge de l'Europe – La Finlande, la Suède et la Norvège face au projet polycentrique européen*, Doctoral thesis, Université de Haute-Bretagne – Rennes 2.

Gløersen, E., Dubois, A., Copus, A. and Schürmann, C. (2006) *Northern Peripheral, Sparsely Populated Regions in the European Union and in Norway*, Nordregio Report (2).

Glückler, J. (2007) Economic Geography and the evolution of networks, *Journal of Economic Geography*, 7, 619–34.

Granovetter, M. (1985) Economic action and social structure: the problem of embeddedness, *American Journal of Sociology*, 91(3), 481–510.

Herdzina, K., Findeis, A., Fleischmann, S.,Wander, C., Piasecki, B. and Rogut, A. (2004) European rural SMEs in the context of globalization and enlargement, in Labrianidis, L. (ed.), *The Future of Europe's Rural Peripheries,* pp. 86–113. Aldershot: Ashgate.

Kaufmann, A. and Tödtling, F. (2000) Systems of Innovation in Traditional Industrial Regions: The Case of Styria in a Comparative Perspective, *Regional Studies*, 34(1), 29–40.

Lindqvist, M., ed. (2010) *Regional Development in the Nordic Countries 2010,* Nordregio Report (2), Stockholm.

Lowe, P.J., Murdoch, J. and Ward, N. (1995) Networks in rural development beyond endogenous and exogenous models, in van der Ploeg, J.D. and van Dijk, G., (eds), *Beyond Modernisation: The impact of endogenous development*. Assen: Van Gorcum.

Marsden, T., Banks, J. and Bristow, G. (2000) Food supply chain approaches: exploring their role in rural development, *Sociologia Ruralis*, 40, 424–38.

Nilsson, B. (2010) Ideology, environment and forced relocation: Kiruna – a town on the move, *European Urban and Regional Studies*, 17(4), 433–42.

Oerlemans, L.A.G. and Meeus, M.T.H. (2005) Do organizational and spatial proximity impact on firm performance?, *Regional Studies*, 39(1), 89–104.

Rosenfeld, S. (2002) *Networks and Clusters: The yin and yang of rural development*, Paper presented at a conference organized by the Centre for the Study of Rural America, 'Exploring policy options for a New Rural America', May 2001. Available at: http://www.kc.frb.org/Publicat/Exploring/RC01Rose.pdf

Ring, J.K., Peredo, A.M. and Chrisman, J.J. (2010) Business networks and economic development in rural communities in the United States, *Entrepreneurship Theory and Practice*, 34(1), 171–95.

Sandow, E. (2008) Commuting behaviour in sparsely populated areas: evidence from northern Sweden, *Journal of Transport Geography*, 16, 14–27.

Swedish Government Office (2007) *En nationell strategi för regional konkurrenskraft, entreprenörskap och sysselsättning 2007–2013 (A national strategy for regional competitiveness, entrepreneurship and employment 2007–2013)*. Available at: http://www.regeringen.se/sb/d/8739/a/74835

Tödtling, F. and Kaufmann, A. (1999) Innovation systems in regions of Europe – a comparative perspective, *European Planning Studies*, 7(6), 699–717.

Uzzi, B. (1996) The Sources and Consequences of Embeddedness for the Economic Performance of Organizations: The Network Effect, *American Sociological Review*, 1996, 61(4), 674–98.

Ward, W.A. and Hite, J.C. (1998) Theory in Rural Development: An Introduction and Overview, *Growth and Change*, 29(3), 245–58.

Young, N. (2006) Distance as a hybrid actor in rural economies, *Journal of Rural Studies*, 22, 253–66.

Young, N. (2010a) Globalization from the edge: A framework for understanding how small and medium-sized firms in the periphery 'go global', *Environment and Planning A*, 42, 838–55.

Young, N. (2010b) Business networks, collaboration and embeddedness in local and extra-local spaces: the case of Port Hardy, Canada, *Sociologia Ruralis*, 50(4), 392–408.

# Chapter 3
# Raising Self-efficacy and Resilience in the Westerkwartier: The Spin-off from Collaborative Leadership

Dirk Roep, Wiebke Wellbrock and Lummina G. Horlings

## Introduction

Place-based approaches to rural development are increasingly favoured (Amin 2004, Healey et al. 2003, Murdoch 2000, Reimer and Markey 2008, Shucksmith 2009, Tomaney 2010). It is argued that they increase self-efficacy through bottom-up development and decentralisation of decision-making (Amin 2004, Böcher 2008, Bruckmeier 2000, Dargan and Shucksmith 2008). In doing so, place-based approaches strengthen the resilience of rural areas against global pressures by decreasing state dependencies and increasing the economic competitiveness of rural areas (Bristow 2010, Taylor 2012). In order to successfully follow through on opportunities for place-based sustainable development, reflexive approaches to governance, replacing hierarchical, policy-centred leaderships with collaborative modes of governance and cross-boundary leadership (Collinge and Gibney 2010), are needed. Enacting collaborative modes of governance requires public support that can relate to the particular assets, challenges and political dynamics of a place (Woods 2007). This must be taken into account in the formulation and implementation of policies at different administrative levels from the EU down to municipalities. The lower the administrative level, the more public policies need to be attuned to the particularities of place.

Grassroots innovations are acknowledged as an important means to deliver sustainability goals (Leach et al. 2012, Wiskerke and Van der Ploeg 2004). The experimental space and support afforded by public policies to bottom-up networks are seen as key to encouraging innovative potential. Public policies can specifically support joint learning, innovation and collective capacities, by enabling networks of grassroots development initiators to develop joint visions, activities and synergies. For this, they need to pool knowledge, capacities and assets, and in doing so they will generate new shared knowledge and capacities (Collinge and Gibney 2010, Gibney 2011). Public policy can facilitate communication and dialogue, partnerships between local and extra-local practitioners, an ethos of social inclusion and structures for democratic decision-making (Collinge and Gibney 2010, Swanson 2001). Building collective agency is thus key to the effectuation of more collaborative modes of governance and, hence, place-based sustainable development (Amin 2004, Gieryn 2000, Healey et al. 2003, Massey 1991, Roep et al. 2009, Swanson 2001). Collective agency results from mutually reinforcing processes of reflexivity and joint capacities built in collaborative activities. Arranging operational interfaces that connect and mediate between the domains of public policies, grassroots development initiatives and knowledge support structures that are supportive to joint learning and innovation in collaborative activities is fundamental to raising collective agency and self-efficacy. This in turn promotes a 'learning rural region' (Wellbrock et al. 2012). Regional reflexivity, or rather place-based reflexivity, defined as the ability of engaged development actors to reflect and to act upon their own activities and interests in order to assess the development of their region and alter their actions accordingly (Gray and Lawrence 2000), will eventually result in effective institutional reform (Roep and Wiskerke 2004), social innovation (Bock 2012, Moulaert et al. 2005) and inclusive development.

In the third part of this chapter this process will be illustrated for the Westerkwartier, a rural area situated to the west of Groningen city in the Netherlands (see Figure 3.1). Here it will be demonstrated how a new collaborative spirit was the impetus for initially informal, and later on formalised collaborations, how collective agency was built and how this was supported by new, well-working arrangements. It is also revealed how this grew in a self-sustaining spiral which enabled a more collaborative mode of governance be developed. Before this discussion, the next section positions the Westerkwartier region within wider political, economic and rural policy developments. In the final section the main issues are drawn together with reflection on wider implications of the case study being made.

**Figure 3.1 Groningen Province and the 'Westerkwartier' (western quarter)**

## The Westerkwartier: Emergent from Collaboration[1]

From a European perspective, the 345km² Westerkwartier is a relatively densely populated (173 inhabitants/ km², about 60.000 in total in 2008), peri-urban area, situated in the western part of Groningen province in the north-west of the

1 This section is based on Jones and Woods (2011) and Wellbrock and Roep (2011), both reports from the EU-funded DERREG research project.

Netherlands. From a Dutch perspective, the Westerkwartier is considered to be a predominantly rural area, due to the dominance (80 per cent) of agricultural land. The Westerkwartier has thus at least two identities which influence its development.

The Westerkwartier comprises the municipalities of Grootegast, Marum and Leek and part of the municipality of Zuidhorn. The area is situated west of the provincial capital Groningen city with its major knowledge centres, public administration offices and regional business centres. The Westerkwartier is well connected to Groningen city and other urban centres by a motorway crossing the area horizontally, railways and a good public transportation system. Consequently, the socio-economic development of the Westerkwartier is predominantly shaped by its interactions with Groningen city and other nearby urban centres. Local businesses are, for example, increasingly focused on providing classic 'rural-fringe' functions and services for Groningen and other close-by cities, including recreational and leisure activities, as well as transport and logistics. Most firms operate in the private sector while employment in the public sector is relatively small. The Westerkwartier has no major industrial employer and business activities include trade, retail and repair businesses. Most of the firms residing in these industrial estates are SMEs with up to 50 employees. Encouraged by strict spatial planning policies, all industrial estates and SMEs are located along the motorway crossing the Westerkwartier or in industrial parks on the fringes of the municipal capitals. Smaller rural firms with major expansion plans also have to relocate to business parks. While remaining in the rural part of the Westerkwartier are micro-firms with fewer than ten employees, these micro-firms are of crucial importance to the vitality of the Westerkwartier and its quality of life, accounting for 73 per cent of private enterprises in the area in 2008. A relatively high level of shops and services has been maintained in the smaller villages and hamlets in the area, making it an attractive place to live in terms of economy and quality of life. Results from a survey among a selection of rural businesses (Copus et al. 2011) affirm that these businesses are strongly anchored in the regional economy and that international trade or network activities have only little direct influences on these rural firms. Nevertheless, the economic vitality of the Westerkwartier is clearly linked to more nationally and internationally oriented businesses. Perhaps this is an explanation why economic growth in the Westerkwartier is above the national average and unemployment in the Westerkwartier is relatively low compared to regional and national averages (as of 2009). The Westerkwartier is thus clearly better positioned economically than areas in the north and east of Groningen province, which, in contrast, experience a shrinking economy, a decline and ageing of population and a subsequent loss of public and private services. It appears to be especially attractive as a residential area for (young) commuters and their families and experiences a net in-migration. A substantial number of Westerkwartier residents, especially from towns and villages around Leek and Marum, commute to work in nearby cities such as Groningen in particular, with newcomers bringing a new interest in the rural area and the quality of life it offers.

The strong division between urban and rural activities enforces a strong sense of rurality in the people of the area. Moreover, most local residents strongly foster its rural identity. The agricultural sector and the industry sector are still seen as the traditional economic pillars of the Westerkwartier, although the contribution of primary agricultural production to GVA and employment has decreased significantly. Nevertheless, in 2008, full-time employment in agriculture was considerably higher in the four municipalities of the Westerkwartier (11 per cent) than on average in Groningen province (3 per cent). Regardless of its relative contribution to the rural economy, agriculture therefore continues to be the economic mainstay of the landscape, and it has become widely recognised that agriculture needs financial support to be able to conserve the landscape as a public good.

Although a fairly small territory, the Westerkwartier is divided in two areas with a distinctive landscape. The rural areas of Grootegast and Zuidhorn municipalities in the northern part are particularly seen as domains of agriculture and designated for that purpose in spatial plans. This part is characterised by an open landscape with large farms and large fields, bordered by alders. The southern part of the Westerkwartier is characterised by a patchwork landscape of small, cultivated fields and pastures bordered by hedgerows. Farms are smaller than in the northern part. The southern part is seen as an attractive residential area with a number of small and large villages in close proximity to larger, urban settlements. This division, reinforced by a motorway and a railway, has a long history and is reflected in a different sense of place to the residents in the two parts.

Spatial plans to protect renowned agricultural landscapes of high cultural and natural value, such as the southern part of the Westerkwartier, have been implemented in the Netherlands since 1975. This has prevented a drastic reconstruction of the area, in particular a resizing of parcels, to further modernise agriculture. Although most farmers opposed a large reconstruction of the area hitherto, there was equal opposition to the demarcation of the area as a protected, valuable agro-cultural landscape. In the first instance, opposition focused on government interference, a situation quickly followed by arguments that protection measures would limit small-scale, farm-based interventions considered necessary for modern farming. In compromise, because of the disadvantages for modern farming practices, farmers were offered payments to compensate income losses and while these were not quickly accepted either, these payments are seen nowadays as a common good by farmers. In addition to these general measures, part of the farmland in the Westerkwartier was destined as a nature conservation area. These parcels were purchased by the Government Service for Land and Water Management and then passed on to the State Forestry Services to manage. This implementation faced strong resistance by farmers in the beginning, but motivated by payments, farmers and the State Forestry Services have recently joined forces to conserve and manage nature and landscape in the Westerkwartier.

Development policies and strategies related to the Westerkwartier argue for a place-based approach which is articulated in the objectives of the 'Development

Plan West-Groningen 2008+' (Groningen 2007). This development plan integrates the national rural development plan 'Agenda Vital Countryside', coordinated by the Ministry of Agriculture, Nature and Food safety (MANF), and the provincial 'Environment Plan Groningen' for the Westerkwartier. The development plan is formulated and implemented by the Steering Group West, a governmental collaboration of the four municipalities, the Groningen Province, the water boards and the involved manager of the 'Agenda Vital countryside'. The plan also integrates the 'LEADER Action Plan 2007–2013' elaborated by the Local Action Group (LAG) of the Westerkwartier, installed in 2007. The development plan for the Westerkwartier follows three broad development lines: increasing economic competitiveness, nature and landscape management and socio-economic vitality. With regard to socio-economic vitality, the 'Development Programme West-Groningen 2008+' in particular acknowledges the importance of small and micro-firms in the rural areas, to the vitality of the Westerkwartier. Since 2008 the LAG supported an increasing number of collaborative initiatives covering activities in various fields such as nature and landscape management, alternative food networks, cultural heritage, rural tourism, renewable energy and new business networks. Being a partnership itself, the LAG supported new collaborations in the Westerkwartier, such as the overarching Westerkwartier Initiative Group, that in turn advocated the installation of a LAG. This will be explained in more depth in the next section.

Despite internal differentiation and rivalry, to better position themselves, the four municipalities in West-Groningen act and present themselves in a unitary way to the outside world. Removing or by-passing obstacles, the Westerkwartier emerges as a vehicle for close collaboration and the promotion of joint interests and shared identity crosscutting public administrative boundaries. With a renowned cultural landscape, minor and major rural settlements and close-by urban centres, specifically the southern part, the Westerkwartier is well equipped and well positioned compared to other less favoured European rural areas. Taken together this provides fertile soil for a fruitful, place-based development.

In the next section, as well as drawing on a considerable amount of secondary data, including work by Wellbrock et al. (2014) and Wellbrock and Roep (2014), the discussion is based on data gathered by personal interviews with various actors in the Westerkwartier area. These include public officers, aldermen, representatives of stakeholder organisations, initiators of various activities and network events, consultants and other professional intermediary persons such as teachers and researchers that have, or have had, an active role in processes that have shaped the region. Part of this process included a workshop where most of the interviewees participated in jointly reflecting on the developments in the area, the role of key actors and events and the way in which these shaped the current position of the Westerkwartier.

## Expanding Collaborative Spirit as Impetus to a Spiral Development of New Collaborations and Supportive Arrangements

From 2003 on, an expanding collaborative spirit became the impetus of joint reflexivity and a growing engagement within the Westerkwartier. Both collaboration and informal collaborative leadership have engendered a spiral of lasting, formal collaborations sustained by arrangements supporting joint reflexivity and joint capacities built in various collaborative activities. This spiral development will be unpacked along two story lines, first the spin-off in terms formal collaborations will be highlighted and then some well-working supportive arrangements will be explained.

The start of the five-year Wageningen UR (University and Research centre) pilot project ‘Brug Toekomst’ (Bridge to the Future) and the parallel rapprochement and cautious collaboration between the two agro-environmental associations in the area coordinating nature and landscape management on farm land and the State Forestry Services commissioned to conserve nature and landscape on former farm land marked an important turn to a more collaborative spirit in the Westerkwartier. The rapprochement, in particular, was quite remarkable due to the long-standing and strong opposition by farmers to designate farm land for nature conservation. However, because of the abilities of the manager of the State Forestry Services to bridge conflicts of interest, to create a sense of common interest and to depict a collaborative vision with regard to nature and landscape management in the Westerkwartier, this rapprochement became possible. Importantly, in a rather informal way, the manager was able to build trust by promoting the common interest and collaboration openly to his superiors and politicians and, more convincingly, by initiating small-scale collaborations. This is where the Wageningen UR pilot project ‘Bridge to the Future’ was significant as it facilitated initial collaborations. The basic idea of the pilot project was to promote joint research and education practices between the university and research institutes and the Van Hall-Larenstein University of Applied Sciences that just had become part of Wageningen UR. A lecturer of Van Hall-Larenstein, and leading member of an agro-environmental association in the Westerkwartier promoting nature and landscape management on farm land, saw the potential and opportunity for students from both Wageningen University and Van Hall-Larenstein being allowed to do assignments commissioned by the agro-environmental association. With his proposal to select the Westerkwartier as a pilot area being granted, the next five years saw 50 students from Van Hall-Larenstein and Wageningen University completing assignments on issues like landscape maintenance and nature development, farming and farm strategies, water management, culture and history, recreation and tourism and liveability and quality of life (Derkzen 2009). The results were presented and discussed with the stakeholders in the Westerkwartier at a yearly regional event (*Streekdag*). This further enhanced dialogue, joint reflection and a collaborative spirit in the region.

The pilot project 'Bridge to the Future' thus brought together an ensemble of committed people living and working in the Westerkwartier that had initiated various activities. In 2004, they united themselves in the *Westerkwartier Streek Initiative* or WSI (Westerkwartier Initiative Group). Through their vision, with support from some enthusiast lecturers and researchers from Wageningen University and Van Hall-Larenstein, the WSI has taken the lead in promoting various collaborative grassroots development activities. This academic support also helped the WSI in facilitating joint reflexivity, setting up participatory activities and establishing a shared vision, as well as to forging strategic alliances and building joint capacities in collaborative projects (Derkzen 2009). The pilot project also helped the WSI to maintain good relations with public administration and in relation to public officers who got involved as a commissioner of student assignments. Municipalities, and with them the province, became willing to partially fund small-scale projects initiated by the WSI. The WSI gradually built fruitful collaborative relations with the four municipalities and the province. This made the WSI ask for additional LEADER funding in 2007. The WSI developed and discussed a LEADER Action Plan 2008–2013 with the initially somewhat reluctant Province of Groningen. The Plan was approved and later, in 2007, a Local Action Group (LAG) of ten members was installed by the Province of Groningen. The LAG included four aldermen each representing one municipality, a secretary from the province and five members recruited from the WSI network. The installation of the LAG thus formalised the already informally established public-private partnership and made funds available for various development initiatives.

At the end of 2007, the 'Bridge to the Future' project expired. Seeing the impact it had, and the rise of a new collective spirit (Derkzen 2009), the WSI wanted to continue and further extend the successful interaction with students and lectures. Consequently, at the end of 2007, a last minute application for financial support was submitted to the Regional Transition Programme (RTP), an innovative education programme approved by the Green Knowledge Cooperative (GKC), of the Van Hall-Larenstein University of Applied Science. The GKC acts as an overarching association of 'green' education institutes in the Netherlands encouraging innovation in 'green' education. The GKC was initiated and funded by the Ministry of Agriculture, Nature and Food Quality (MANF) governing 'green' education and research in the Netherlands. The RTP promotes the *Kenniswerkplaats* (Knowledge Atelier), an innovative methodology in green education by means of which students and lecturers get engaged in investigating regional questions commissioned by regional stakeholders (Foorthuis and Lutz 2013). The idea is to create a strong learning environment for students, generating practical and sometimes unconventional perspectives and stimulating contributions to regional development. Since this methodology was quite similar to what was developed in the 'Bridge to the Future' project, RTP financial support was granted for one year. In September 2008, the Westerkwartier was visited by a group of policymakers, researchers and the funding scheme coordinator. They were exploring the wider implementation of the Knowledge Workplace, were interested to hear about the experiences and lessons learned and assessed whether there was broad support for setting up a Knowledge Workplace in the Westerkwartier (Roep et al. 2009). In particular the political (programme) and financial (co-funding) support of the municipalities and province was seen as crucial for considering the Westerkwartier as one of the 15 pilot regions for implementing and further developing the Knowledge Workplace methodology in the Netherlands. By the end of 2010, the Westerkwartier was finally selected as a pilot region with funds available for operating a Knowledge Workplace and developing a joint vision and regional development programme as the basis for structural collaboration. This started a further widening of the collaboration. In spring 2011 a 'masterclass' started to discuss and elaborate a joint vision and work programme and, in the end, some formal arrangements. This process was facilitated by a team from the Knowledge Workplace programme, including a member of the WSI well acquainted with green education. In autumn 2011 a provisional programme was presented and the idea of a Westerkwartier Cooperative was launched to implement the programme and coordinate the activities. It lasted until the end of 2012 before the Westerkwartier Cooperative was officially launched.

In about nine years the initial informal collaborative network of the WSI evolved in a step-by-step manner into a successive range of productive formal collaborations thus effectuating a collaborative mode of governance (Collinge and Gibney 2010). The founders of the WSI displayed and excelled as informal collaborative leaders able to meditate well between leaders of grassroots initiatives, public officers, politicians, lecturers, researchers and consultants. They were the spider in a web of relations (Nyhan 2007), acting as an interface between the domains of public administration, knowledge support structure and the rural area. They actively promoted what Wellbrock et al. (2012) framed as a 'learning rural region' or a 'learning rural area' (Wellbrock et al. 2013). Their ability to get things done, first little by little and then increasingly by acquiring additional funds, made their efforts successful. The WSI, with political and financial support from the LAG, were important for generating wider engagement in the area. This was crucial to the spiral development of expanding collaborative spirit and rewarding collaborations in the area. In this respect the international Countryside Exchange in 2008, the instalment of the Rural House in October 2008 and a number of Rural Café's in the course of 2009 and 2010 appeared to be particularly successful arrangements (Wellbrock et al. 2011).

The Countryside Exchange (CE) is a method in which an international team of experts pays a well-prepared visit to a rural area, consults its residents and organisations, analyses issues and obstacles, assesses resources and abilities and discusses opportunities. The aim is to mobilise residents and stakeholder organisations in a rural area, have them jointly appreciate their assets and capacities and reflect on future developments, inspire them and present a list of viable activities as a follow-up of the visit and eventually stimulate collaboration (Adviesgroep 2009, Derkzen 2009, 2008). The issues and questions raised by the hosting area are leading and prepared in advance of the visit. The findings and recommendations of the expert team are presented and discussed at a closing symposium. The CE in the Westerkwartier was set up in collaboration with the European Network for Local Development Teams (ENLDT). The ENLDT is rooted in LEADER and Local Action Groups (LAG) to encourage mutual, cross-cultural learning among practitioners, local governments and experts. Both visitors and hosts mutually benefit from these visits. The CE was initiated by the WSI and LAG and co-funded by MANF and a regional bank. A Dutch consultancy group and the already active researchers helped prepare the CE and facilitated the visit. As reported (Adviesgroep 2009, Derkzen 2009), the CE came at the right moment, soon after the instalment of the LAG. The CE proved to be very effective in raising self-esteem, energising and mobilising people, extending networks, creating a collaborative spirit and vision and encouraging regional learning (Derkzen 2009). The CE had a significant impact as it enabled the WSI and LAG to further specify and substantiate future development activities and to take the next step in the formalisation of public-private collaborations.

In October 2008 the Rural House was set up. This was an initiative of the WSI and LAG, funded by LEADER with co-financing from the municipalities and province. The Rural House is located in a house in the main shopping street of Grootegast, in the middle of the Westerkwartier. The main idea here was to offer a low threshold entry for residents with rural development related questions and grassroots initiatives. 'They should see it as their place, not as an outpost of the public administration' as one of the founding members remarked. Next to offering a physical meeting place for the WSI and other assemblies, the Rural House offered a physical and virtual window for consultation with a team of experts from different organisations. Next to public officers of the municipalities and province the team consisted of experts in rural and regional development policies and supporting measures, experts from DLG and the *Vereniging Groninger Dorpen* (VGD; Association Groningen Villages) and the earlier mentioned touristic catalysts. The DLG is a state agency in charge of EU-payments and active in land consolidation, land acquisition and the design and realisation of nature and landscape plans. The VGD promotes the interest of (small) villages, but also offers professional consultancy to related rural development initiatives and networks. The expert team pooled their expertise and divided the support asked for among the group. The expert team was seen as very useful and adequate by the beneficiaries (Wellbrock and Roep 2011), not only in helping out with applications for subsidies, but also for their brokering capacities, alongside their ability to bridge and connect. The expert team operated well as an interface between grassroots initiatives and policy schemes and, in some cases, public knowledge institutes and private agents when additional expertise was needed. Thus the expert team actively promoted the learning rural area. Above all, the Rural House provided a low threshold access to support and expertise needed in grassroots development initiatives. The success of the expert team depended a great deal on the operational flexibility assigned by the LAG in the approval of applications for subsidies to small-scale initiatives such as setting up networks, organising events and minor investments of rural entrepreneurs. The expert team was authorised to grant subsidies up to a maximum of €10,000. This operational flexibility contributed to the well-approved decisiveness of the team.

The Rural Café was also an initiative of the WSI and LAG and funded by LEADER. The idea here was to create an informal, low threshold gathering to enable all residents to meet, initiate and discuss ideas and projects and be inspired by short presentations. In 2009 and 2010 four such Rural Cafés were organised around a theme to ensure that people with a shared interest could meet and interact. The Rural Cafés were well attended, up to 100 people, and well appreciated. In effect it mobilised and inspired a lot of people resulting in wider support and new initiatives.

Next to these long-term arrangements to generate wide participation and enhance joint reflexivity, short-term targeted operational support for specific collaborative initiatives was also arranged. These arrangements were spin-offs from the new collaborative ethos that emerged, and created new collaborations themselves. Two such arrangements, of short duration, were regarded as particularly successful by its beneficiaries, the supporting professionals as well as the LAG that initiated and funded these arrangements (Wellbrock and Roep 2011). One was the installation of three so-called touristic catalysts to foster networking and collaboration in rural tourism. Rural tourism was identified as a sector with growth potential by the LAG which recognised the lack of collaborative entrepreneurial spirit in the area and consequently targeted it for support. As the alderman of Zuidhorn, chair of the LAG at the time, explained in 2009:

> We have 200 firms in the tourism and leisure sector, but all of them work on their own, without any joint vision. They are not part of a business network and do not get any financial public support to develop their business potential. To be eligible for financial support we asked them to collaborate. We used this as an extra incentive.

Three professionals were thus assigned for a period of three years to foster collaborative entrepreneurship in rural tourism. The arrangement proved to be quite rewarding because of the professional skills of these catalysts and the operational flexibility granted to them (Roep et al. 2011). Their effort resulted in a self-sustaining platform for rural tourism and eventually a business association.

Another fruitful arrangement was the short-term assignment of a professional of the VDG (*Vereniging Dorpen Groningen*, Association Villages Groningen) to facilitate *Wichterwest*, a new born network of business women. The VGD was founded in the 1970s to defend public services in smaller villages against a provincial plan to reduce public services in the countryside. Today, the VDG offers professional support and grants funds for liveability projects in villages. The VDG promotes the establishment of community houses, helps make village visions and plans strong enough to apply for funding and acts as facilitators toward grassroots development initiatives. The VGD has thus developed into a well-working operational interface, *Wichterwest*, well equipped to initiate and facilitate. Professionals of the VGD assisted start-up business women to make a business plan, organise network meetings and obtain additional funding for their activities from the province of Groningen. *Wichterwest* has developed into a support network itself, organising meetings and workshops to facilitate exchange and learning and provide expertise for women starting in a business career. Both arrangements are examples of successful operational interfaces which stimulated collaboration and raised entrepreneurial skills. In both cases the assigned professionals operated as intermediary agents or brokers with specific sets of tasks. Next to their professional skills the allowable operational flexibility appeared to be key to their productive operations (Roep et al. 2011). Within the limits of the tasks assigned they could decide for themselves what could be done best in consultation with the beneficiaries.

## Collaborative Leadership, Collective Agency and Effective Institutional Reform: Key Ingredients for Raising Self-efficacy and Resilience

The case of the Westerkwartier is an example of how effective institutional reform can evolve from a process of joint, place-based reflexivity. It started in a rather informal way, a group of visionary people, in this case assembled around a pilot project, engaging with place-based development in an area. In the Westerkwartier, this gave momentum to a new collaborative spirit, involving the engagement of residents and various activities, thereby gradually building joint capacities to make things work according to their shared values and vision. This collective agency was consolidated step by step in various new institutional arrangements. The case thus demonstrates the importance of arranging well-working interfaces (Wellbrock and Roep 2011) between (a) bottom-up development initiatives, (b) supporting policy schemes and (c) facilitators of joint learning and innovation coming from education, research and advice. These operational interfaces further catalysed and actively promoted a 'learning rural area' (Wellbrock and Roep 2011). In addition to the integration of various policy schemes into a single incentive programme for the area, these arrangements created the experimental space and support for bottom-up networks that are seen as crucial for grassroots innovations to flourish (Leach et al. 2012, Roep and Wiskerke 2004).

The extent and robustness of the achievements in the long term remain to be seen. Those involved, however, experienced substantial progress, but at the same time they acknowledged that what was achieved so far was still fragile. In an earlier stage, those involved in the spiral development already identified key ingredients for fruitful collaborations (Derkzen 2009, Traa 2011): informal leadership and vision; initiative and self-efficacy; involvement and commitment; mutual confidence; joint learning; smart combinations of themes and partners; long-lasting alliances and new roles and structures. Basically, these observations all point to the same source: a need for collaborative leadership, building collective agency as a result of a process of joint learning by doing, and new arrangements to create a more favourable institutional setting.

Sotarauta et al. (2012) sustain this conclusion based on the analysis of a diversity of cases: collaborative leaders can be institutional entrepreneurs turning external stimuli into internal changes. They can act as animators who bring to life, enliven, spark, create and produce, showing a strong commitment to build collective agency and reform the institutional setting. As part of joint or social learning processes, different stakes, domains and administrative levels can be bridged and connected, creating new knowledge by sharing and drawing together explicit and tacit knowledge across institutional, disciplinary and sectorial boundaries (Wals 2007). This is not only a matter of strategy, but also a process of building trust, creating commitment and of reframing different values and knowledge sets into a more coherent, collaborative narrative. Sol et al. (2012) picture trust, commitment and reframing as emergent properties of interaction in their account of social learning in the earlier mentioned pilot project 'Bridge to the Future'. Institutional entrepreneurs, being 'actors who have an interest in particular institutional arrangements and who leverage resources to create new institutions or to transform existing ones' (Garud et al. 2007, p. 957), go beyond existing frames of reference, to facilitate experimentation and the exploration of innovative, networked approaches for resilience (Sotarauta et al. 2012, pp. 489–90).

The Westerkwartier case revealed key ingredients for raising self-efficacy in shaping more resilient places. But are these ingredients the only thing? Is there a single recipe for making rural areas more resilient? If not, what is so special about the Westerkwartier? The Westerkwartier as a bounded territorial entity with a shared sense of place and common identity emerged only recently, as a result of the spiral development painted above. This development is still in progress and has not yet stabilised itself. When compared to, for instance, nearby rural areas in the east of Groningen Province, or to some of the case study areas, such as Alytus County in Lithuania or Comarca de Verin in Spain and Regierungsbezirk Dresden in Germany as discussed in other chapters in this book, the Westerkwartier made significant progress (Roep et al. 2011, Wellbrock et al. 2013). The case study area in Saarland (Germany), in contrast, resembled the Westerkwartier (Wellbrock et al. 2013). A comparative analysis reveals important differences that mark a dissimilar starting point and can explain the differential progress in terms of the spiral development made in the different areas. First of all, differences in political histories echo in actual political dynamics and in the prevailing mode of governance. The prevailing mode of governance can be quite different, from a policy-centred mode to a more collaborative mode. Secondly, the prevailing mode of governance is in turn reflected in the institutional setting and influences whether the institutional setting is more or less favourable to effectuate place-based development strategies. The less favourable the institutional setting is, the less fertile ground there is for a spiral development to flourish, the more difficult it is to make progress and the more collaborative, visionary leaders are needed to initiate it.

The case study areas Alytus County and Regierungsbezirk Dresden were, for instance, part of transition countries. They faced radical institutional changes or had to deal with institutional voids. These circumstances mark quite a difference to the situation in the Westerkwartier. In addition, people out-migrate from Alytus County and Regierungsbezirk Dresden and these areas do not attract newcomers, as is the case in the Westerkwartier. Some areas are clearly better equipped in terms of valuable resources, whereby less favourable conditions might be demotivating for residents to engage in place-based development. In these circumstances, building collective agency to shape a more resilient place appears to require much more effort to initiate than in the Westerkwartier.

Further, the comparison shows that the way in which places were shaped in the past matters for shaping more resilient places today, including their position within wider (re)structuring forces that affect and shape them and their connections to other places. Collaborative leadership, collective agency and effective institutional reform are identified as key ingredients for raising self-efficacy and resilience, but their spiral development and potential to shape resilient, sustainable places has to be situated against the backdrop of the wider picture. This has been done in Figure 3.2, framing the spiral development in shaping of resilient places at the intersection of unbound political-economic, socio-cultural and ecological structuring processes (the blades). From a relational perspective, place is framed as a time- and space-bounded expression of this threefold of intersecting processes, resulting from the interaction and co-evolution of human and non-human agency. Landscape is a clear tangible expression and renders an identity to a rural area. Boundaries like identities are, however, constructions, they evolve and can be contested, as in the case of the Westerkwartier. Through the threefold of unbound processes, the shaping of each place is somehow interconnected with the shaping of other places and part of a mutually affecting 'worldwide' web of interconnections (Massey 2004) and flows of natural resources, capital, ideas, technology, goods and people. These interconnections and flows reflect patterns of domination and differentiation between places (Escobar 2008). These unbound forces are mediated by human and non-human agents and agencies and act as catalysts (Woods 2013). They are reflected in arrangements and in institutional settings that further

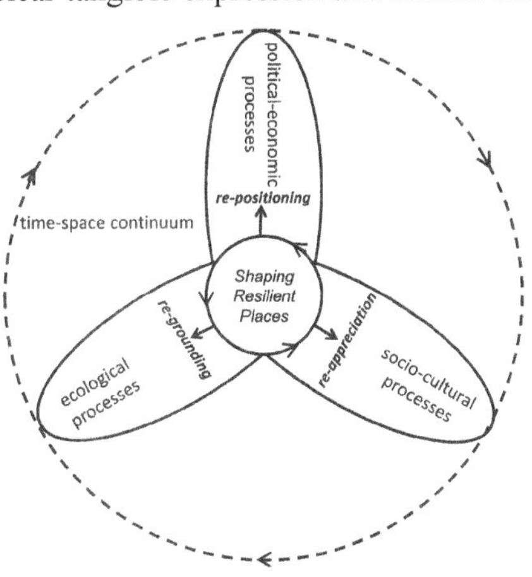


**Figure 3.2 Shaping resilient places at the intersection of unbound political-economic, socio-cultural and ecological structuring processes: reappreciation, regrounding and repositioning resulting from raised, place-based self-efficacy**

shape interactions and interconnections, such as policy schemes formulated and implemented at different administrative levels, just as much as in technological arrangements and or market arrangements.

The story of the spiral development can now be framed accordingly. A raised joint reflexivity, collaborative spirit and collective agency, inspired by collaborative leadership and sustained by supportive arrangements, has had a spin-off in formal collaborations and effective institutional reform. This spiral evolved by a *reappreciation* of the bonds and binds that place entwines with a *regrounding* of development activities in place-based resources that enforce the capacity of residents to *reposition* their place towards more hegemonic unbound forces and renegotiate the conditions of engagement. This indeed marks an increased self-efficacy and hence, as assumed, resilience in the long run.

## Acknowledgement

We are grateful to those people in the Westerkwartier that gave of their time and shared their knowledge during the course of this research.

## References

Adviesgroep, E. (2009) *Countryside Exchange*. Available at: http://www.countryside-exchange.nl/index.php?option=com_content&task=view&id=29&Itemid=38 [Accessed 30 November 2009].

Amin, A. (2004) Regions Unbound: Towards a New Politics of Place, *Geografiska Annaler. Series B, Human Geography*, 86(1), 33–44.

Barca, F. (2009) Pursuing equity through place-based development policies. Rationale and the equity efficiency issue, *OECD/TDPC Symposium on Regional Policy*, Paris.

Böcher, M. (2008) Regional Governance and Rural Development in Germany: the Implementation of LEADER+, *Sociologia Ruralis*, 48(4), 372–88.

Bock, B.B. (2012) Social innovation and sustainability; how to disentangle the buzzword and its application in the fi eld of agriculture and rural development, *Studies in Agricultural Economics*, 114, 57–63.

Bristow, G. (2010) Resilient regions: re-'place'ing regional competitiveness, *Cambridge Journal of Regions, Economy and Society*, 3(1), 153–67.

Bruckmeier, K. (2000) Leader in Germany and the Discourse of Autonomous Regional Development, *Sociologia Ruralis*, 40(2), 219–27.

Collinge, C. and Gibney, J. (2010) Connecting place, policy and leadership, *Policy Studies*, 31(4), 379–91.

Copus, A., Dubois, A., Hedström, M., Kairyte, E., Stastna, M., Slavič, I. and Wellbrock, W. (2011) *DERREG D1.4 Summary of Research Findings*. Stockholm, Sweden.

Dargan, L. and Shucksmith, M. (2008) LEADER and Innovation, *Sociologia Ruralis*, 48(3), 274–91.

Derkzen, P. (2008) *Countryside Exchange Westerkwartier*, Unpublished: Report for the European Network for Local Development Teams (ENLDT). Unpublished.

Derkzen, P. (2009) *Geleerd in het Westerkwartier. Een onderwijsproject in een dynamische regio 2003–200*. Wageningen: Wageningen University.

Escobar, A. (2008) *Territories of Difference. Place, movements, life, redes*. USA: Duke University Press.

Foorthuis, W. and Lutz, S. (2013) *Kenniswerkplaats as a Method for Learning Regions and Lifelong Learning*. Groningen: AOC Terra.

Garud, R., Hardy, C. and Maguire, S. (2007) Institutional entrepreneurship as embedded agency: an introduction to the Special Issue, *Organization Studies*, 28, 957–69.

Gibney, J. (2011) Knowledge in a 'Shared and Interdependent World': Implications for a Progressive Leadership of Cities and Regions, *European Planning Studies*, 19(4), 613–27.

Gieryn, T.F. (2000) A Space for Place in Sociology, *Annual Reviews of Sociology*, 26, 463–96.

Gray, I. and Lawrence, G. (2000) Reflexivity and Neoliberalism in Regional Australia, in *Sociological Sites/Sights, TASA 2000 Cconference*, Adelaide, Australia, December 6–8, Flinders University.

Groningen, S.W. (2007) *Ontwikkelingsplan regio West Groningen 2008+*. Groningen, NL: Provincie Groningen.

Healey, P., de Magalhaes, C., Madanipour, A. and Pendlebury, J. (2003) Places, identity and local politics: analysing initiatives in deliberative governance in Hajer, M. and Wagenaar, H. (eds) *Deliberative Policy Analysis. Understanding governance in the network society*, pp. 60–86. Cambridge: Cambridge University Press.

Jones, L. and Woods, M. (2011) *D5.2 Westerkwartier Contextual Report*, Aberystwyth, UK. Available at: Aberystwyth University: http://www.derreg.eu/content/resource-centre/d52c-westerkwartier-contextual-report

Leach, M., Rockström, J., Raskin, P., Scoones, I., Stirling, A. C., Smith, A., Thompson, J., Millstone, E., Ely, A., Arond, E., Folke, C. and Olsson, P. (2012) Transforming Innovation for Sustainability, *Ecology and Society*, 17(2 C7–11).

Lowe, P., Murdoch, J. and Ward, N. (1995) Networks in Rural Development: Beyond Exogenous and Endogenous Models, in Ploeg, J.D. v.d. and Dijk, G.v. (eds) *Beyond Modernization: the impact of endogenous rural development*, pp. 87–106. Assen, The Netherlands: van Gorcum.

Marsden, T. and Bristow, G. (2000) Progressing Integrated Rural Development: A Framework for Assessing the Integrative Potential of Sectoral Policies, *Regional Studies*, 34(5), 455–69.

Massey, D. (1991) A Global Sense of Place, *Marxism Today*, 24–9.

Massey, D. (2004) Geographies of Responsibility, *Geografiska Annaler. Series B, Human Geography*, 86(1), 5–18.
Moulaert, F., Martinelli, F., Swyngedouw, E. and Gonzalez, S. (2005) Towards Alternative Model(s) of Local Innovation, *Urban Studies*, 42(11), 1969–90.
Murdoch, J. (2000) Networks – a new paradigm of rural development?, *Journal of Rural Studies*, 16(4), 407–19.
Nienaber, B. (2007) Die Rolle der Landwirtschaft im Konzept der Eigenständigen Regionalentwicklung – geographisch untersucht am Beispiel der Region West Irlands, *Münstersche Geographische Arbeiten*, 50.
Nyhan, B. (2007) Building learning regions for innovation in Europe: a challenge for education and training, in Gustavsen, B., Nyhan, B. and Ennals, R. (eds) *Learning Together for Innovation: promoting learning regions*, pp. 16–34. Luxembourg: Office for Official Publications of the European Communities.
O'Brian, K. (2011) Responding to environmental change: A new age for human geography?, *Progress in Human Geography*, 35(4), 542–9.
OECD (2006) *The New Rural Paradigm: Policies and governance*. Paris: OECD Publishing.
Ray, C. (2006) Neo-endogenous development in the EU, in Cloke, P., Marsden, T.K. and Mooney, P. H. (eds) *Handbook of Rural Studies*, pp. 278–91. London: SAGE Publications Ltd.
Reimer, B. and Markey, S. (2008) *Place-based Policy: A rural perspective*. Canada. Available at: http://www.crcresearch.org/files-crcresearch_v2/ReimerMarkeyRuralPlaceBasedPolicySummaryPaper20081107.pdf
Roep, D., Horlings, I. and Wielinga, E. (2009) *De werkvloer van een Kennisnetwerk Vitaal Platteland. Kennis maken met regionale kennisarrangementen*, Den Haag, the Netherlands: LEI Wageningen UR. Available at: http://www.lei.dlo.nl/publicaties/PDF/2009/2009–049.pdf
Roep, D., Wellbrock, W., Kairyte, E., Farrell, M., Mahon, M., Frys, W., Nienaber, B., Kriszan, M., Nadler, R., Burdack, J. and Dominguez Garcia, L. (2011) *DERREG D4.4 Summary Report*, Wageningen, The Netherlands. Available at: http://www.derreg.eu/content/resource-centre
Roep, D. and Wiskerke, J.S.C. (2004) Epilogue: Reflecting on novelty production and niche management in agriculture, in Wiskerke, J.S.C. and Ploeg, J.D. v.d. (eds) *Seeds of Transition: Essays on novelty production, niches and regimes in agriculture*, pp. 341–56. Assen: Royal van Gorcum.
Shucksmith, M. (2009) Disintegrated Rural Development? Neo-endogenous Rural Development, Planning and Place-Shaping in Diffused Power Contexts, *Sociologia Ruralis*, 50(1), 1–14.
Sol, J., Beers, P.J. and Wals, A.E.J. (2012) Social learning in regional innovation networks: trust, commitment and reframing as emergent properties of interaction, *Journal of Cleaner Production*.
Sotarauta, M., Horlings, I. and Liddle, J. (2012) Knowledge leadership for resilient regions: Concluding remarks, in Sotarauta, M., Horlings, I. and Liddle, J. (eds)

*Leadership and Change in Sustainable Regional Development*, pp. 271–90. Abingdon: Routledge.

Swanson, L.E. (2001) Rural Policy and Direct Local Participation: Democracy, Inclusiveness, Collective Agency, and Locality-Based Policy, *Rural Sociology*, 66(1), 1–21.

Taylor, B.M. (2012) Regionalism as resistance: Governance and identity in Western Australia's Wheatbelt, *Geoforum*, 43(3), 507–17.

Tomaney, J. (2010) *Place-based Approaches to Regional Development: Global trends and Australian implications*, Sydney, Australia: Australian Business Association. Available at: http://www.abfoundation.com.au/research_project_files/52/Tomaney_Rev.Final_Colour.pdf

Traa, F. (2011) *Regional Learning Arrangements in the Westerkwartier*, 12–13th October 2011, Murska Sobota, Slovenia. Available at: http://www.derreg.eu/content/events/final-conference-derreg-project

Wals, A.E.J. (2007) *Social Learning Towards a Sustainable World. Principles, perspectives and praxis*. Wageningen: Wageningen Academic Press.

Wellbrock, W., Nienaber, B., Kriszan, M., Nadler, R., Roep, D. and Frys, W. (2013) Arranging support of collective learning and innovation in German rural regions in Tamasy, C. and Revilla Diez, J. (eds) *Regional Resilience, Economy and Society*, pp. 195–212. Farnham: Ashgate.

Wellbrock, W. and Roep, D. (2014) The Learning Rural Area Framework: A Heuristic Tool to Investigate Institutional Arrangements which Support Collaboration in Rural Areas. *Sociologia Ruralis*, available online 10 June 2014. DOI: 10.1111/soru.12049

Wellbrock, W. and Roep, D. (2011) *Deliverable 5.3 'Westerkwartier'; Case study summary report*, Wageningen, The Netherlands: Wageningen University. Available at: http://www.derreg.eu/content/resource-centre

Wellbrock, W., Roep, D., Kairyte, E., Farrell, M., Mahon, M., McDonagh, J., Frys, W., Nienaber, B., Kriszan, M., Nadler, R., Burdack, J. and Dominguez Garcia, L. (2011) *DERREG Deliverable 4.3: Summary of good practise examples: Well-working arrangements for support of joint learning and innovation in Europe's rural regions*. Wageningen, The Netherlands. Available at: Wageningen University: http://www.derreg.eu/content/resource-centre

Wellbrock, W., Roep, D., Mahon, M., Kairytė, E., Nienaber, B., Dominguez Garcia, M. D., Kriszan, M., and Farrell, M. (2013) Arranging public support to unfold collaborative modes of governance in rural areas, *Journal of Rural Studies*, 32, 420–429.

Wellbrock, W., Roep, D. and Wiskerke, J.S.C. (2012) An integrated perspective on rural regional learning, *European Countryside*, 4(1), 1–16.

Wiskerke, J.S.C. and Van der Ploeg, J.D. (2004) *Seeds of Transition: Essays on novelty production, niches and regimes in agriculture*. Assen: Royal Van Gorcum.

Woods, M. (2007) Engaging the global countryside: globalization, hybridity and the reconstitution of rural place, *Progress in Human Geography*, 31(4), 485–507.

Woods, M. (2013) Rural development, globalization and European regional policy: perspectives from the DERREG project, *Geographica Polonica*, 86–110.

## Chapter 4

# Managing Development with Civil Society in a Globalizing Rural

Marie Mahon, Maura Farrell and John McDonagh

### Introduction

This chapter builds on earlier discussions in this volume about the ways in which effective knowledge and learning interfaces are instrumental in determining rural futures under globalization. Using examples from the DERREG West of Ireland case study, it situates the discussion within the DERREG interpretative (transformationalist) framework on globalization (Woods 2009), and employs the heuristic model devised by Wellbrock et al. (Chapter 3) which explores the significance of knowledge and learning interfaces as key to the dynamic of rural localities. It draws on the concept of civil society to further explain the nature of these relations between locally based groups and institutional actors at various scales in a development context. Because rural civil society represents a highly heterogeneous and increasingly complex realm, the kinds of involvement with development have been similarly varied. In this latter regard, a focus on the institutional arrangements of different civil society organizations provides potential insights into local development priorities and the capacity to implement these as innovative and sustainable responses to globalization in specific place-based contexts. Drawing on Woods' (2007) assertions about the significance of the micro-scale as the key site of renegotiation and reconstitution of the rural under globalization, it suggests that a more critical perspective on identifying civil society in the rural would provide informed insights on its particular range of perceptions and expectations of globally connected rural places, on the expectations for its role in their continued development and sustainability and on how this development role would be best arranged within national and EU policy and programme requirements.

### Civil Society in a Globalizing Rural; a Diverse Sphere in a Multifaceted Development Domain

Recent developments in rural studies have emphasized how a relational perspective on the reconstitution of rural places under globalization requires spatially and contextually grounded explanations of this global interconnectedness to make sense of how it impacts on rural places and societies (Woods 2007). Identifying and interpreting locally specific experiences of globalization are bound up in the multiple dimensions to globalization – cultural, political, social and economic (ibid.). This notion of globalization as a dynamic multidimensional and transformationalist process connects with wide-ranging perspectives on civil society as a highly complex phenomenon conditioned by these same contexts from which it emerges, with globalization being identified as exerting specific transformative influences. Exploring civil society as a means to understand the dynamic of contemporary rural society and place under globalization means identifying in the first instance what individuals, groups and activities this now encompasses.

Since the 1970s, the concept of civil society, which has accounted broadly for the sphere of citizen interaction with the state and the market, has become part of the discourse of international development policy as a means to build societal resilience against a range of development challenges (Inglehart 1990, UNDP 1993, World Bank 1997). The various philosophical positions on civil society, from 'societas civilis' to bourgeois, activist, neoliberal and postmodern conceptions, reflect a broader set of perspectives about its capacity to function as a societal realm within which a range of engagements between citizens, the state and the markets takes place (Hyden 1997, Kaldor 2003, McElwaine 2007). McElwaine (1998) organizes these perspectives into two broad groupings: (a) the liberal approach, viewing civil society as a largely autonomous sphere of freedom and liberty, stressing the beneficial effects of active civic associations (de Tocqueville); (b) the neo- or post-Marxist approach, viewing civil society as a site of oppression and power inequalities (Marx, Hegel, Gramsci). During this period, ambiguity surrounding civil society as a political goal or project and as functioning reality (often in the form of an NGO), which might not measure up to that goal, has left it open to sustained criticism, especially in relation to its role as a political project that resists or influences the state and the market (McElwaine 1998, 2007; Bebbington and Bebbington 2001, Escobar 2001, Mercer 2002, Mohan 2002, Bebbington 2004). Edwards (2009) argues that this has been mainly to do with a failure to provide clarity on its meaning, the trend to apply it strategically (for example in a power context, to promote government agendas) and a preoccupation with its form as voluntary associations. His reminder is that there are necessarily different interpretations and dimensions of civil society which must be critiqued as such, but not as separate strands; civil society needs to be understood as a form of collective action, functioning within specific but collective contexts. As such it can also be understood as inherently geographical. Wagner (2012) similarly critiques slippages in application of the concept, focusing particularly on the consequences for interpreting the role of different organizations and groups. Through an analysis of institutional arrangements, he distinguishes between economic, political and hybrid organizations, indicating a potentially wide range of groups and associated activities distinguishable through criteria such as organizational principles, governance systems, the role of volunteers and the type of collective action involved. This argument for greater precision in distinguishing between groups in order to establish clearer insights into their relative positions on local development reflects other critiques, particularly those by Raco (2005) on sustainable development as actual place-based development practice constituted by competing rationalities and resulting hybrid approaches involving government and civil society.

Keane (2003) refers to the emergence of highly heterogeneous and complex development relationships with the market and the state as a marker of civil society's particular ideological and policy positions on globalization (for example neoliberalization, corporatization, social capital/social self-organization or activism) (Anheier 2007).

Among the strongest critiques of the relationship between civil society and the state in contemporary development debates are those from a regulationist perspective that connects 'rolled out' forms of neoliberalism with increasingly interventionist forms of governance and regulation that facilitate capital accumulation (Jessop 2002, Raco 2005). The neoliberal state is seen as having played to a growing tension between political civil society and the activities of NGOs as its 'organized' version in terms of whether it holds advocacy or welfare provision roles (Herbert-Cheshire 2000, Cheshire 2006). This version of civil society reflects specific aspects of liberal philosophy that endorse associational activity and a strong voluntary sector (Putnam 1993, 2000). Its representative forms, including the voluntary sector and not-for profit-organizations, increasingly fill the economic and social needs gap arising from uneven processes of capitalist development and reduced levels of direct state intervention. Third sector and social economy values are endorsed as part of neocommunitarian strategies to promote local economic development, harnessed through local groups' ability to mobilize local resources in line with state and market interests (Gerometta et al. 2005). Civil society as part of the wider project of neoliberalism becomes transformed into a compensatory mechanism for the inadequacies of the market (Jessop 2002). In the UK context, it is argued that the effective mainstreaming of the third sector into public policy has brought about a fundamental reconfiguring of state-third sector relations organized largely on government terms (Dahrendorf 2001, Fyfe 2005). Codes and strategies associated with implementation of state-third sector partnerships have been linked to Foucauldian ideas of new governmental rationalities that scale control upwards (Swyngedouw 2000). On a global stage, Kaldor (2003) questions the likelihood of challenge to the state or the markets by such 'tamed NGOs' that are essentially working to 'smooth the path of economic globalization' (p. 11).

## Study Location and Methodology

The specific focus of this discussion is County Roscommon, part of the Border Midland West Regional Authority Area. The county is predominantly rural, with a population of 64,000 (CSO 2011); most of it had been under the remit of the now discontinued CLÁR Programme (Ceantair Laga Árd-Riachtanais), which funded regeneration measures in rural areas dealing with population decline. The county suffers from significant infrastructural deficiencies although it boasts a wealth of other under-exploited resources in the form of an unspoilt natural landscape, spectacular natural amenities such as lakes, rivers and forestry and a rich historical and cultural legacy. The largest town, Roscommon, has a population of almost 5,700 (CSO 2011). In the context of this chapter, four case studies involving local grassroots organizations were identified as examples of rich and diverse interfaces within the local rural development domain that featured a range of actors and organizations that facilitated learning and knowledge exchange (Wellbrock et al. 2012). Semi-structured interviews and focus groups were conducted with 22 representatives of grassroots representatives (GR) (as representatives of civil society), and with 11 relevant local and regional state organization and institutional representatives (IR). These interviews focused on the nature of cooperation and the challenges for achieving sustainable rural development at the local and regional level particularly in the increasingly global context of rural change. Among the issues explored was the degree of empowerment and autonomy believed to be available at the regional and local level to maximize on government and other institutional supports, including those that related to rural development programmes. The interviews were analysed using the learning rural area interpretative framework devised by Wellbrock et al. (2013) to map, analyse and compare how public policies support various interfaces through which a range of development actors learn to work together. The following discussion initially gives brief critique of the prevailing model of governance underpinning rural development as a context to the interview discussion and the analysis of institutional space provided for those grassroots civil society organizations involved in rural development activities. It identifies and interprets the nature and strength of interfaces with other grassroots organizations and with institutional actors at different scales of cooperation. It then provides an analysis of the nature of resulting interfaces and arrangements between various institutional actors; how in the context of specific place-based conditions and circumstances agency is exerted and collective action is facilitated; how much of this is achieved at the local level among civil society groups; how specific globalizing phenomena make a difference to the process of rural development and how these are negotiated.

## Critiquing Institutional Arrangements; Civil Society's Role in Local Rural Development

Hadenius and Uggla (1996) offer an extensive analysis of the forms of relationships civil society can have with the state, and the power of the state to facilitate or hinder civil society as an active pluralist sphere. In this regard they view the state as ultimately the holder of certain common resources required to promote the growth of civil society. The state is also instrumental in providing facilitative structures for civil society to engage in governance as one route to participating in development. Notwithstanding Ireland's adoption of a neo-corporatist partnership model for national development which can be linked to the EU's requirement for area-based participatory approaches to delivering programmes such as LEADER (Geddes and Le Galès 2001, McCafferty and Walsh 1999), its tendency to retain centralized control of key decision-making in governance arrangements is noted (Adshead 2005). The concept of participatory governance has resonated to the extent that the local development landscape has over the last two decades or so come to be more densely populated with institutional groups and agencies; however, this has not necessarily offered the institutional scope for these groups to operate with civil society organizations to influence the direction of development discourses or take account of globalizing trends and influences.

The creation of organizations such as Pobal that answer directly to government rather than allocate these functions to regional or local administrations, and the limited authority ceded to umbrella groups such as the City and County Development Boards (CDBs), are more reflective of increased authoritarian than democratizing processes of governance (Adshead 2005). Ó Broin and Waters' (2007) review of local governance structures, particularly public agencies and their relationship with regional, national and EU levels of governance, suggests a significant democratic deficit between the state and civil society in terms of lack of autonomy in decision-making around public expenditure, management and planning on the part of local and regional organizations, and limited ability to respond in a flexible and context-specific way to development needs (Laffan 1996, Breathnach 2002). Knickel et al. (2008) argue that institutional arrangements committed to delivering endogenous and integrated rural development must be able to facilitate agencies as reflexive organizations at the micro-scale, to reduce coordination problems and enhance cooperation in rural development. From a transformative perspective (Woods 2007), this also means a legitimate institutional focus on place-specific, locally relevant social constructions of rurality and locally identified rural development needs; that is, managing a process of capacity building for a broader-based locally relevant perspective on development rather than project-based interventions which in themselves can be exclusionary and devisive from a civil society perspective (Geddes 2000, Geddes and Le Galés 2001). The impact from a globalizing standpoint is in the possible negative consequences from weakly developed and fragmented local civil society that has not advanced some collective consciousness of the development opportunities that globalization might present, or how it may wish to resist certain aspects of it.

As a layer of governance, local government in Ireland has traditionally had little involvement in social and economic policy discussion or implementation. In 2000, Strategic Policy Committees (SPCs) as a form of new partnership arrangement with local and voluntary community sectors and other representative bodies were put in place under the umbrella of CDBs. At the same time, the previous direct governance arrangements that area-based partnerships for social inclusion had with the European Commission became mediated by central government via a separate agency – Area Development Management, later known as Pobal. McCafferty and Walsh (1999) discuss the significance of the partnership model as a means to enhance community-level involvement in development, and promote more innovative approaches to local level intervention. Callanan's (2005) research on the SPC process of co-opting identified interests into institutional structures highlights the potential advantages for improved cooperation between local government representatives and civil society in relation to development at this local level. In their comparative cross-border study of voluntary and community sector involvement with public administration in development, Acheson and Williamson (2007) suggest, however, that the CDB institutional structure has had the effect of weakening the community and voluntary sectors by drawing them into policy and programme coordination instead of supporting their capacity to engage in more direct framing and provision of programmes. As institutional structures, CDBs have neither executive authority nor access to programme funding that might empower groups through local-level interventions, or through being able to significantly extend themselves beyond local arenas into wider networks of global influence.

In Ireland, LAGs deliver both LEADER and the Local Development Social Inclusion Programme. LAGs are viewed as key organizations in terms of how they network between stakeholders at all governance scales. They maintain a strong link to central government due to the mainstreaming of LEADER as part of EU rural development policy, and because of their reliance on government for matched funding. In 2009, the Irish government initiated a consolidation of LAGs which involved reducing the overall number and loosely aligning them to local authority administrative boundaries, that is, one per county with the major urban locations also retaining one. In rural counties, single LAGs now deliver both LEADER and social exclusion programmes. Dwyer and Maye (2009) have suggested that this combined remit has promoted a uniform model of Axis 3 (quality of life and rural economic diversification) that aligns more to mainstream social exclusion measures, but that diverts from the LEADER approach (Axis 4) which promotes area-based, innovative actions. Since late 2011, the Minister for the Environment, Community and Local Government confirmed arrangements for bringing LAGs under the remit of county and city councils. This is in spite of concerns being expressed openly by LAGs via national and local media about the potential undermining of their current capacity to operate innovatively, which is already felt to be overly conditional on central government and EU managerial and auditing constraints. The same concerns were articulated by several institutional interviewees who felt it worked against the fundamental bottom-up development ethos of the LEADER initiative, which would cease to be perceived as an entity dedicated to local rural social and economic development (IR 7, 9, 10). Boyle (2000) has referred to the legacy of central state control over local and regional government in terms of a traditionally undeveloped local-scale dependency, the incentive to connect to these lower scales of governance for their existence having never been a priority. At the other end of the state-civil society spectrum, under the new proposals for incorporation of LAGs under local government structures, the separate and more specific place-based identity of these groups risks being eroded (GR 1, 2, 11); their new identity as part of local government also presents certain risks for their status as civil society groups in terms of potential compromises about their remit and the perception in the eyes of the local communities they aim to represent.

Institutional representatives described an increasingly disempowered sphere of local governance driven mainly by decisions emanating from central government level. Several raised the issue of reductions in funding and staffing, lack of information on the future of development programmes and budgets, limited scope for decision-making and disproportionate reporting requirements as creating difficult environments for local development. One representative involved with local groups (IR 7) referred to the transfer in control of development programmes from one government department to another following on from Ministerial reshuffles or national parliamentary elections. These changes frequently meant transfer of the programmes to alternative agencies, often with a changed remit and associated eligibility and reporting requirements, with little or no consultation with local groups who continued to make efforts to respond and adapt to these requirements. This was particularly the case where groups were themselves in primarily employment-based service-delivery roles as opposed to volunteering organizations. In this regard, the direction of rural development was also directly connected to their own employment or enterprise futures and concerns were raised by several local group representatives who proposed changes in policy or institutional strategies would put these in danger (GR 1, 2, 3, 4, 6, 7). In other words, a sense of heightened vulnerability rather than empowerment was the predominant sentiment at these levels.

On the surface, the partnership model has allowed local organizations, and through these, civil society, to be acknowledged as legitimate actors in regional and rural development. The observation made by one institutional representative (IR 1) was that local and regional organizations in particular had been able to develop more regional and local level policy-relevant knowledge and administrative expertise in response to EU and central government funding conditions, something that has been identified as a major strength of regional-level governance (Moylan 2009). Another remarked that through the conditions attaching to Structural Funds, strong local-level networks of action groups and partnerships have been facilitated via a succession of social funding and rural development programmes (IR 5). In other words, a level of regional and local capacity building had taken place which could be interpreted as reflecting a strong set of knowledge and learning interfaces between institutional and civil society actors at a certain scale of interaction. However, the continued emphasis on market-based notions of competitiveness, with associated development paths for regional and rural development has led to specific mechanisms for realizing these. In development terms, funding interventions such as those that are part of the National Development Plan, and associated programmes such as LEADER, all come with strict time and budgetary limits, as well as specific and quite strictly defined programme objectives. These were also identified as tending to drive development agendas, with much of the energy of local and institutional groups diverted into administrative and reporting tasks set from the top down (IR 7, 9, 10, 11: GR 1, 2, 3, 7, 10, 11). The emphasis is inevitably upon funding measurable outputs rather than on sustaining processes such as cooperation between organizations or groups at the local level (Hadenius and Uggla 1996).

## The Extent of Rural Identities in the Realm of Rural Civil Society

The way that rural civil society features in a development capacity has also been influenced by the continued association with mainstream agriculture. This dominant perception of agriculture as the driver of rural economic growth and sustainability is regarded as a defining factor in central government's approach to rural policy (Varley et al. 2009). In the Irish context, farmers' representatives comprise a specific group under the partnership model. The proportion of households still engaged in farming is significant (29,750 full-time and 69,451 part-time farms) (Teagasc National Farm Survey 2010). In 2010, direct payments under CAP accounted for €1.8 billion of total farm payments, which in real terms accounts for up to 95 per cent of farm incomes (ibid. 2010). Agriculture has been identified by Matthews (2011) as the single most substantial contributor to the economic and social viability of rural areas, and farming groups remain the most prominent and influential representative bodies for rural populations. However, Hubbard and Gorton (2011) suggest that the contribution of agriculture to rural economies throughout the EU is static or even in decline, with other forms of rural activity and rural networks coming to the fore in sustainability terms. This predominant agricultural sector representation through the partnership model generates a potentially disproportionate perspective on what are much more diverse contemporary rural populations and on who constitute the active agents of rural civil society (Mahon 2007, Woodward and Halfacree 2002).

This dominance of agricultural concerns has been reflected in Ireland's Rural Development Programme (2007–13) which placed the maximum allowed funding (80 per cent) into Axis 2 (Environment and countryside) for agri-environmental and related schemes. The remainder was split between Axis 1 (Competitiveness of farming and forestry) and Axis 3 (Quality of life and rural economic diversification). Axis 4 dealt with the LEADER Approach. Maye (2008, p. 3) sums up Ireland's RDP as 'primarily a plan to address environmental priorities', with farmers and land/agriculture-related concerns forming the primary focus. Responsibility for Axes 1 and 2 remained predominantly under the remit of government departments dealing with mainstream agriculture, while departmental responsibility for Axes 3 and 4 had changed at least twice, from agriculture to Community, Rural and Gaeltacht Affairs (DCRGA), and in 2011 to Environment, Community and Local

Government (DECLG). The different departmental responsibility for Axes 3 and 4 arguably signified a concern to emphasize more social and cultural aspects to rural development. These featured under the NDP's Rural Social & Economic Development Programme. It comprises four subprogrammes including: LEADER/Rural Economy; Rural Social Scheme; Western Investment Fund; CLÁR (Ceantair Laga Árd-Riachtanais – regeneration of rural areas with population decline). A further NDP priority area, Social Inclusion, also featured several small-scale programmes (a number of which have also changed departmental ownership at least once) that encompass rural areas. These were administered by Pobal, usually via LAGs. In the overall development funding context, the amounts available to these subprogrammes were modest (see also Lowe and Ward 2007).

The fragmented nature of the programmes for rural development that were principally administered by LAGs in terms of their spread across several government departments and the relative frequency (within the lifetime of the NDP) with which they transferred between departments was reported by interviewees to have absorbed time and resources from both LAGs and grassroots organizations involved in development (GR 1, 2, 3, 7, 10, 11, IR 7, 10, 11). Revised programme parameters were among the most time-consuming issues, because they required new learning and readaptation by groups. Other operational costs such as printing new information leaflets or signage were to reflect the title of the government department in which the programme now resided. Disruption to networks of cooperation built up over time between personnel in government departments, agencies and local groups, and the effort and time required to form new sets of connections and working relationships were also part of these essentially top-down reorganization processes. Reflecting the hierarchical nature of funding control and a predominant patron/client relationship with government, changed programme remits meant local groups having to alter the core focus (and original strength) of local projects to ensure a continued stream of funding.

One of the arguments advanced in favour of Pobal's role has been that a single agency would more effectively manage programme delivery between local groups and government. However, since Ireland's economic crisis, Pobal's own resources had been reduced, and it no longer had the same capacity to act as a supporting interface to engage in the process of development through mentoring local groups to build capacity as actors in local development. Its own internal structure (a board appointed by government, representing community, voluntary and business interests) has no representation at regional or local levels. In a power context, the nature of the organization's relationship with central government is essentially a contractual one. The effect of funding cut-backs had begun to similarly impact on other statutory organizations that provided mentoring supports to local groups, particularly the inability of agency staff to spend time on the ground, visiting local initiatives, discussing with them their specific needs, exchanging information, connecting them with other agencies or initiatives and providing targeted advice and support.

In the Roscommon case study area, networking arrangements between agencies and organizations were well-established at the local level, and between the local and the regional level. From a development perspective, this cooperation largely reflected an informal level of local governance, designed to overcome fragmentation of effort and maximize on limited resources. Agencies were represented on one another's boards of directors, project evaluation committees or other subcommittees, and there was good interchange of local information on development needs and pressures. Although these were important components of the process of development intervention, it was also apparent that their effectiveness was curtailed by an inability to operate more independently at the local level and to make decisions in a reflexive, context-specific way. For example, narrow categories set by central government identifying potential projects by scale gave little real priority to area-based or innovative development potential.

LAGs have aimed to reflect a broad community representation through the composition of their boards of management, and the evidence in this research suggested that this layer of governance was highly proactive and engaged in the delivery both of funded programmes on behalf of government, and in identifying ongoing local development needs (Kovách 2000, 2006). The composition of boards of local agencies has come in for certain criticism in terms of their representativeness both of agencies and the community (Curtin and Varley 2002, Shortall 2004, Varley and Curtin 2006); efforts to balance this include time and frequency limits to terms and positions of office. All of the local public agencies were also represented on these boards, in turn enabling LAGs to be part of dense networks of local institutional interactions. The benefits for community representatives were that they had access to information about funding or other opportunities, or to operational issues such as change in programmes, which constituted a range of additional resources around which they could make informed and timely decisions about their own development strategies. As a form of social capital, local groups clearly realized certain benefits from being part of these institutional networks. From a wider perspective, however, the advantages could be construed as limited, particularly when the local agencies and central government are themselves operating under increasingly constrained circumstances. The argument has been strongly made in the context of community development that the strength of social capital lies in the value of resources to which access can be realized (DeFilippis 2001) and the level of real empowerment that this bestows (Knickel et al. 2008).

## Concluding Remarks

Building on the discussions introduced by Roep et al. in this volume, this chapter has sought to develop further perspectives into the status of civil society in development agendas for a globalizing rural through its involvement with existing institutional structures and arrangements to promote rural development. It first reviewed the concept of civil society from a philosophical standpoint to illustrate the extent of various debates on the meaning and role of civil society, and how these become linked to different perspectives on globalization and development. It then considered its role in a rural development context, especially the way this has been shaped by EU and national rural development and sustainability agendas, but also by an increasingly globalized development context. The emphasis on multi-level governance and partnerships with civil society (Marks and Hooghe 1996, Bache and Flinders 2004) has seen more evidence of regulation and control of development than a sharing of decision-making and competences. Even at the meso-level where opportunities for engagement in development are brought somewhat closer administratively and geographically to the local level, the parameters for reflexive actions are constrained and depend on the innovation and creativity of agencies working with local groups to harness development opportunities.

In a rural development context, bottom-up development initiatives which explicitly embrace civil society groups are endorsed to the extent that they are written in to macro-level policy and programme strategies (Knickel et al. 2008); however, whether civil society groups find themselves in an empowered and emancipatory institutional space that can constantly challenge and inform development policy and process, or whether a rhetoric of empowerment that comes with bottom-up discourses has kept them as locally relevant programme implementation bodies is less clear. In spite of high levels of professionalization and competence acquired by civil society groups in the course of over two decades of Structural and other funding supports, their involvement in programme delivery has continued to be on a contractual, project-based footing which has at the same time successfully delivered several rounds of LEADER and social inclusion programmes. Although there is an extensive governance framework in place that underpins development interventions at a range of spatial scales and involving a series of organizations and agencies, this has not translated into a regionally or locally specific approach to policy formulation or programme delivery which would provide competitive advantage to rural locations that might benefit from, or be vulnerable to globalizing processes. Civil society groups that operate outside of this remit, for example that have established themselves on a more voluntary footing, occupy an essential sphere of activity in rural community life which yields important social and also economic benefits; however their perspective is less focused on competitive gains for rural areas which is where the issue of globalizing forces is of most concern.

The implications for civil society seeking to operate at this more competitive globalizing seam but in more constrained policy and institutional circumstances are not clear. The predominant institutional relationship, that is, centralized control of resources and focus on outputs rather than process (Hadenius and Uggla 1996, Jones 2001, High and Nemes 2007, Allen 2009) seems unchanging with the forthcoming consolidation of partnership and LEADER companies under the remit of local government. The concern here is of the LEADER structure becoming absorbed into the institutional space of local government to the point that it becomes indistinguishable from it and, as such, becomes distanced from its original civil society associations. A counter-argument advanced by central government is that a move to bring groups under the local authority umbrella represents a shift towards more representative forms of local governance. For many members of civil society, the notion of local government in Ireland as a democratic or accountable forum in the wake of recent spatial planning and other administrative crises lacks credibility. These new arrangements for rural development in Ireland open up both threats and opportunities for rural civil society in the form of more innovative forms of rural civil society partnerships that are in many ways more free to chart their own development course in a globalizing context. To this end, the 'anarchic element' to rural development first suggested by Ray (2000) may find scope to emerge.

## References

Acheson, N.V. and Williamson, A.P. (2007) Civil society in multi-level public policy: the case of Ireland's two jurisdictions, *Policy & Politics*, 35 (1), 25–44.

Adshead, M. (2005) Europeanization and changing patterns of governance in Ireland, *Public Administration,* 83(1), 159–78.

Allen, J. (2009) Three spaces of power: territory, networks, plus a topological twist in the tale of domination and authority, *Journal of Power*, 2(2), 197–212.

Anheier, H. (2007) Reflections on the concept and measurement of global civil society, *Voluntas*, 18, 1–15.

Bache, I. and Flinders, M. (2004) *Multi-level Governance*. Oxford: Oxford University Press.

Bebbington, A. and Bebbington, D. (2001) Development alternatives: practice, dilemmas and theory, *Area*, 33, 7–17.

Bebbington, A. (2004) NGOs and uneven development: geographies of development intervention, *Progress in Human Geography*, 28(6), 725–45.

Boyle, M. (2000) Euro-regionalism and struggles over scales of governance: the politics of Ireland's regionalisation approach to Structural Fund allocations 2000–2006, *Political Geography*, 19, 737–69.

Breathnach, P. (2002) Regional government: the missing link in the National Spatial Strategy? Paper presented to the Regional Studies Association (Irish Branch) National Conference 'Ireland 2020: People, Place and Space', 26 April, Bunratty, Co. Clare, Ireland.

Callanan, M. (2005) Institutionalizing participation and governance? New participative structures in local government in Ireland, *Public Administration*, 83(4), 909–29.

Central Statistics Office (2011) *Census of population classified by area.* Dublin: Central Statistics Office.

Cheshire, L. (2006) *Governing Rural Development: Discourses and practices of self-help in Australian rural policy*. Aldershot: Ashgate.

Curtin, C., and Varley, T. (2002) Changing patterns of leadership and local power in rural Ireland, in Halfacree, K., Kovách, I. and Woodward, R. (eds) *Leadership and Local Power in European Rural Development*, 12–32. Aldershot: Ashgate.

Dahrendorf, R. (2001) *Challenges to the voluntary sector. Arnold Goodman Lecture 17th July 2001*. Tonbridge: Charities Aid Foundation.

DeFilippis, J. (2001) The myth of social capital in community development, *Housing Policy Debate*, 12(4), 781–806.

Dwyer, J. and Maye, D. (2009) *Assessing the second round of CAP pillar 2 programmes – insights from policy design, delivery and context*. Paper delivered to the Joint AESI/AES Conference, 30 March, Dublin, Ireland.

Edwards, M. (2009) *Civil Society*. Cambridge: Polity Press.

Escobar, A. (2001) Culture sits in places: reflections on globalization and subaltern strategies of localization, *Political Geography*, 20, 139–74.

Fyfe, N.R. (2005) Making space for 'Neo-communitarianism'? The third sector, state and civil society in the UK, *Antipode*, 37(3), 536–57.

Geddes, M. (2000) Tackling social exclusion in the European Union? The limits to the new orthodoxy of local partnership, *International Journal of Urban and Regional Research*, 24(4), 782–800.

Geddes, M. and Le Galès, P. (2001) Local partnerships, welfare regimes and local governance: a process of regime restructuring? in Geddes, M. and Benington, J. (eds) *Local Partnerships and Social Exclusion in the European Union: New forms of local social governance?* London: Routledge.

Gerometta, J., Häusermann, H. and Longo, G. (2005) Social innovation and civil society in urban governance: Strategies for an inclusive city, *Urban Studies*, 42(11), 2007–21.

Government of Ireland (2007) *National Development Plan 2007–2013. Transforming Ireland: A better quality of life for all*. Dublin: Government Publications Office.

Hadenius, A. and Uggla, F. (1996) Making civil society work, promoting democratic development: What can states and donors do? *World Development*, 24(10), 1621–39.

Herbert-Cheshire, L. (2000) Contemporary strategies for rural community development in Australia: a governmentality perspective, *Journal of Rural Studies*, 16, 203–15.

High, C., and Nemes, G. (2007) Social learning in LEADER: Exogenous, endogenous and hybrid evaluation in rural development, *Sociologial Ruralis*, 47(2), 103–19.

Hubbard, C. and Gorton, G. (2011) Placing agriculture within rural development: evidence from EU case studies, *Environment and Planning C: Government and Policy*, 29, 80–95.

Hyden, G. (1997) Civil society, social capital, and development: Dissection of a complex discourse, *Studies in Comparative International Development*, 32(1), 3–30.

Inglehart, R. (1990) *Culture Shift in Advanced Industrial Societies*. Princeton, NJ: Princeton University Press.

Jessop, B. (2002) Liberalism, neoliberalism, and urban governance: A state-theoretical perspective, *Antipode*, 34(3), 452–72.

Jones, M. (2001) The rise of the regional state in economic governance: 'partnerships for prosperity' or new scales of state power? *Environment and Planning A*, 33, 1185–211.

Kaldor, M. (2003) *Global Civil Society; An answer to war*. UK: Polity Press.

Keane, J. (2003) *Global Civil Society?* Cambridge: Cambridge University Press.

Knickel, K., Schiller, S., Vihinen, H. and Weber, A. (2008) New institutional arrangements in rural development, in van der Ploeg, J.D., and Marsden, T. (eds) *Unfolding Webs: The dynamics of regional rural development*, pp. 111–28. Netherlands: van Gorcum.

Kovách, I. (2000) LEADER, a new social order, and the Central-and East-European countries, *Sociologia Ruralis*, 40(2), 181–89.

Kovách, I. and Kuĉerová, E. (2006) The project class in Central Europe: The Czech and Hungarian cases, *Sociologia Ruralis*, 46(1), 4–21.

Laffan, B. (1996) Ireland: a region without regions – the odd man out? in Hooghe, L. (ed.), *Cohesion Policy and European Integration: Building multi-level governance*. Oxford: Oxford University Press.

Lowe, P. and Ward, N. (2007) Sustainable rural economies: some lessons from the English experience, *Sustainable Development*, 15, 307–17.

Mahon, M. (2007) New populations; shifting expectations: the changing experience of Irish rural space and place, *Journal of Rural Studies*, 23, 345–56.

Marks, G. and Hooghe, L. (1996). European integration from the 1980s: state-centric v. multi-level governance, *Journal of Common Market Studies*, 34 (3), 341–78.

Matthews, A. (2011) Agriculture, rural development and food safety, in O'Hagan, J. and Newman, C. (eds), *The Economy of Ireland: National and sectoral policy issues*, pp. 215–42. Dublin: Gill & Macmillan.

Maye, D. (2008) National report on RD policy design – Ireland; Assessing the impact of rural development policies (incl. LEADER) (RuDI, FP7). Deliverable D 2.2.

McCafferty, D. and Walsh, J. (1999) *Local partnerships in local governance: the sub regional dimension of devolution*. Paper presented to the European Regional Studies Association 39th European Congress, 23–27 August, Dublin, Ireland.

McElwaine, C. (1998) Civil society and development geography, *Progress in Human Geography*, 22(3), 415–24.

McElwaine, C. (2007) From local to global to transnational civil society: Re-framing development perspectives on the non-state sector, *Geography Compass*, 1(6), 1252–81.

Mercer, C. (2002) NGOs, civil society and democratization: a critical review of the literature, *Progress in Development Studies*, 2, 5–22.
Mohan, G. (2002) The disappointments of civil society: the politics of NGO intervention in Northern Ghana, *Political Geography*, 21, 125–54.
Moylan, K. (2009) *The future of EU Cohesion Policy and its implications for Irish regional policy*. Paper presented to the Regional Science Association International Conference, 3rd September, Limerick, Ireland.
O'Broin, D. and Waters, E. (2007) *Governing Below the Centre: Local governance in Ireland*. Dublin: New Island Books.
Oireachtas Committee on Jobs, Enterprise and Innovation (2013) *Rural Communities Report: Discussion with Minister for the Environment, Community and Local Government, Tuesday, 12th February*. Dublin: Houses of the Oireachtas.
Ostrom, E. (1990) *Governing the Commons: The evolution of institutions for collective action*. Cambridge: Cambridge University Press.
Putnam, R. (1993) *Making Democracy Work: Civil traditions in modern Italy*. Princeton, NJ: Princeton University Press.
Putnam, R.D. (2000) *Bowling Alone: The collapse and revival of American Community*. New York: Simon & Schuster.
Raco, M. (2005) Sustainable development, rolled-out neoliberalism and sustainable communities, *Antipode*, 37(2), 324–47.
Ray, C. (2000) The EU LEADER programme: Rural development laboratory, *Sociologia Ruralis*, 40(2), 163–71.
Shortall, S. (2004) Social or economic goals, civic inclusion or exclusion? An analysis of rural development theory and practice, *Sociologia Ruralis*, 44(1), 109–23.
Shucksmith, M. (2009) Disintegrated rural development? Neo-endogenous rural development, planning and place-shaping in diffused power contexts, *Sociologia Ruralis*, 50(10), 1–14.
Swyngedouw, E. (2000) Authoritarian governance, power, and the politics of rescaling, *Environment and Planning D: Society and Space*, 18, 63–76.
Teagasc (2010) *National Farm Survey*. Agricultural Economics & Farm Surveys Department, Teagasc, Ireland.
United Nations Development Programme (1993) *UNDP and organisations of civil society: building sustainable partnerships*. New York: United Nations.
Urry, J. (2000) *Sociology Beyond Societies: Mobilities for the twenty-first century*. London: Routledge.
Van der Ploeg, J.D., van Broekhuizen, R., Brunori, G., Sonnino, R., Knickel, K., Tisenkopfs, T. and Ooostindie, H. (2008) Towards a framework for understanding regional rural development, in van der Ploeg, J.D., and Marsden, T. (eds) *Unfolding Webs: The dynamics of regional rural development*, 1–28. Netherlands: van Gorcum.
Van Rooy, A. (2008) Civil society as idea: an analytical hat stand, in Van Rooy, A. (ed.) *Civil Society and the Aid Industry*, 6–30. London: Earthscan.

Varley T. and Curtin, C. (2006) The politics of empowerment: power, populism and partnership in rural Ireland, *The Economic and Social Review*, 37 (3), 423–46.

Varley, T., McDonagh, J. and Shorthall, S. (2009) The politics of rural sustainability, in McDonagh, J., Varley, T and Shorthall, S. (eds), *A Living Countryside? The politics of sustainable development in rural Ireland*, pp. 1–22. Aldershot: Ashgate.

Wagner, A. (2012) 'Third sector' and/or 'civil society': a critical discourse about scholarship relating to intermediate organisations, *Voluntary Sector Review*, 3(3), 299–328.

Wellbrock, W., Roep, D. and Wiskerke, J.S.C. (2012) An integrated perspective on rural regional learning, *European Countryside*, 4, 1–16.

Wellbrock, W., Roep, D., Mahon, M., Kairyte, E., Nienaber, B., Domínguez García, M.D., Kriszan, M. and Farrell, M. (2013) Arranging public support to unfold collaborative modes of governance in rural areas, *Journal of Rural Studies*, 32, 420–429.

Woodward, R. and Halfacree, K. (2002) Influences on leadership and local power in rural Britain, in Halfacree, K., Kovách, I. and Woodward, R. (eds) *Leadership and Local Power in European Rural Development*, 59–90. Aldershot: Ashgate.

Woods, M. (2007) Engaging the global countryside: globalization, hybridity and the reconstitution of rural place, *Progress in Human Geography*, 31(4), 485–507.

Woods, M. (2009) *Developing Europe's Rural Regions in the Era of Globalisation: An interpretative model for better anticipating and responding to challenges for regional development in an evolving international context. DERREG FP7 Deliverable 5.1 Literature Review and Conceptual Framework.*

World Bank (1997) *World Development Report 1997: The state in a changing world*. Oxford: Oxford University Press.

# Chapter 5
# Local Cultures of the Ceredigion Economy: Practices of Endogenous Development in Rural Mid Wales

Laura Jones, Jesse Heley and Suzie Watkin

## Introduction: Placing Rural Mid Wales

Ceredigion is a rural country located on the west coast of the British Isles, where, flanked by the Irish Sea to the west and the Cambrian Mountains to the east, and with poor internal and external transportation links, it is conventionally regarded as occupying a peripheral economic location within Wales, Britain and the European Union (EU). In the latter context, it lies within the 'West Wales and the Valleys' NUTS2 statistical region. This follows the vertical division of Wales into West and East based on identified differences in regional development potentials and needs, shaped over time, and, in part, by spatial proximity to the English border and the apparent economic advantages this affords in terms of transport linkages and access to markets. Although East Wales contains the large rural county of Powys it is regarded as a 'Competitive and Employment' region. West Wales and the Valleys, on the other hand, is one of only three UK regions which currently fall under the EU's Convergence Objective for the 2007–13 round of structural funds due to low average wages and low levels of economic activity – the others being Cornwall and the Isles of Scilly, and the Scottish Highlands and Islands (European Commission 2009). Prior to this, Ceredigion was the recipient of Objective 1 funding (2000–06) for 'lagging behind' regions.

At the same time, West Wales and the Valleys is itself socially and economically diverse. The EU classifies the region as 'intermediate rural' as it takes in a diversity of landscapes; extending from the urban areas of Swansea, Bridgend and Port Talbot in the south, and the former industrial heartland of the South Wales Valleys, through an expanse of predominantly rural (mixed agricultural, mountainous and coastal) terrain across the 'Mid Wales' counties of Pembrokeshire, Carmarthen and Ceredigion, before rising into Gwynedd and the mountains of Snowdonia in the north and across the Menai Straits to the Isle of Anglesey. Consequently, the region experiences a diversity of social and economic development opportunities and needs stemming from, amongst other factors, the decline of mining industries (coal, metal ores and slate) and ongoing agricultural transformation. As such, the European Union's ambition to make West Wales and the Valleys 'an entrepreneurial region at the cutting edge of sustainable development' (European Commission 2011) is, and will continue to be, pursued in dialogue with specific local contexts and cultures, needs and wants. In this chapter, we focus within this broader region on the coastal Mid Wales local authority of Ceredigion, where rural isolation and peripherality remain structural factors to be negotiated alongside multifaceted globalization processes, in determining Ceredigion's political-economic development trajectory within the emergent global countryside.

Covering an area of 1,795 sq km and with a population of 76,938 (2010), Ceredigion has the second lowest population density in Wales (Source: Data Unit Wales). Large areas of the Ceredigion landscape are afforded protected status due to their exceptional environmental qualities, including sections of the 100-km-long coastline between the Dyfi and Teifi estuaries and the wild, sparsely populated Cambrian Mountain range. The settlement pattern of the county is characterized by dispersed small towns, villages and farms, with only three towns having a resident population of more than 2,000; these being Aberystwyth (20,000 approx), Cardigan (4,000 approx) and Lampeter (3,000 approx). This topography, coupled with poor communication links, present particular challenges in terms of service provision and also business growth and sustainability, with restricted access to skilled labour markets and customer catchment areas.

Economically, the historic centrality of agriculture in Ceredigion has been somewhat overtaken in recent decades by the growth of the service sector. The locality's main economic driver is now the administrative and service centre of Aberystwyth where already high levels of public sector employment are further supplemented by the relocation of several Welsh Government departments to the town in 2010. The universities at Aberystwyth and Lampeter, and the National Library of Wales at Aberystwyth, are important local employers as well as national cultural institutions. Indeed, Ceredigion together with other largely rural counties such as Gwynedd and Anglesey are considered the traditional Welsh language heartlands, with 52 per cent of the Ceredigion population recorded as Welsh speakers in 2009; roughly double the Wales average (Welsh Government 2011).

Consequently, Ceredigion has the potential to provide insights into the ongoing processes of economic restructuring and differentiation occurring across the Welsh countryside (see Day et al. 1989, Milbourne 2012), as the declining dominance of agricultural employment has given rise to new modes of economic, social and political relations (Marsden et al. 1993, Marsden 1998). This is not to suggest that agriculture is no longer significant, as it continues to be not only an important economic practice but also occupies a central role in social and cultural understandings of place. Rather, Ceredigion (and rural Wales more broadly) as an economic space is becoming increasingly diverse, as advances in ICT infrastructure (in some areas) and personal mobility (for some people) are perceived as challenging the constraints of a 'space-based' economy and relative distance from urban cores; constraints which earlier top-down (modernization) approaches to rural development sought to overcome through inward investment, industrialization and improvements in transport infrastructure. The 'new paradigm' of place-based development discourse identifies 'the capacity of a rural locality to sell itself as an attractive place for economic investment' as key to its economic prosperity; emphasizing endogenous investment, bottom-up innovation and territorially integrated development, framed within a broader paradigm of sustainable development (Woods 2009, p. 187; and see Cheshire 2006, van der Ploeg et al. 2002). However, this requires a multiscalar approach which, on the one hand, emphasizes local embeddedness and place-specific resources, knowledges and cultural forms, whilst also looking outwards in order to sell this place-specificity to extra-local (national, international) consumers. This policy direction is not without risks, however, with the ability to mobilize local actors a key factor in the relative 'success' or 'failure' of different regions in social and economic terms (Barca 2009).

This chapter considers examples of endogenous economic development from Ceredigion, where the importance of place is attested to through business practices focused upon selling 'localness' to a global market, and where 'place-branding' is used to tell particular stories about the provenance of food, drink and clothing, as well as experiences in eco- and luxury tourism. These types of business practice provide a means of responding to the obvious challenge of competition with other rural localities, whereby relative peripherality becomes a lifestyle factor and selling point in and of itself. Nonetheless, living and doing business within such peripherality has its trade-offs, and we highlight the persistence of structural disadvantage (geographical position, transport and communications infrastructure) as a crucial factor determining relative prosperity and the economic development trajectories pursued by local actors in Ceredigion.

It is clear that low population limits local trade, and physical isolation, whilst embraced by some parts of the tourist industry, can restrict the potential for market expansion. Concurrently, such structural limits to endogenous development can be repositioned as positive factors for the smaller scale development ambitions and survival of locally owned and run enterprises, whereby there is a degree of protection and insularity afforded from the highs and lows of global financial processes and from new vulnerabilities introduced through the integration of rural economies into global networks. In this way, we consider how more predictable trajectories based around stability and lifestyle are prioritized by some local economic actors, wherein environmental and social goals can be integrated with economic ones.

Unlike other chapters in this volume, which are based on research for the DERREG project, this chapter draws primarily on research conducted as part of the Wales Institute for Social and Economic Research, Data and Methods (WISERD) Locality Programme, funded by the Higher Education Funding Council for Wales (HEFCW) and the UK Economic and Social Research Council (ESRC). In particular, the chapter draws on documentary analysis and interviews with key stakeholders undertaken to examine the structures and dynamics of social and economic governance in the region, and on interviews with local businesses conducted to investigate the impact of the economic recession. The chapter also draws on research as part of the Wales Rural Observatory on migrant workers in rural Wales, funded by the EU Rural Development Programme for Wales, which involved statistical analysis and survey interviews with migrant workers from Poland, Portugal and the Philippines.

## Engaging the Global – Past and Present

Whilst Ceredigion might be considered representative of rural Wales[1] to an extent, it also experiences particular local factors that condition the way globalization trends are playing out *in situ*; from historical linkages to the wider world through livestock-droving journeys and mining exports, to contemporary factors, including the impact of Ceredigion's higher education institutions and their increasing numbers of overseas students. For example, Aberystwyth University's 2011 intake included students from 69 countries (the top ten being Poland, Malaysia, China, Bulgaria, Norway, Ireland, India and the USA). The county also has a long-standing political culture of liberalism, which, allied to the settled intelligentsia attached to the universities, was a notable factor in encouraging significant 'back-to-the-land' in-migration during the latter half of the twentieth century (see Halfacree 2007). This political context has furthermore arguably contributed towards shaping a strong contemporary environmental ethos, manifested in support for the 'Transition Towns' movement and other forms of green living.

In this way, we follow the imperative laid out by this volume's editors, and others previously (see Murdoch 2006, Woods 2007), to move beyond generalist accounts of globalization as a homogenizing force, towards the place-based study of globalization processes as they are negotiated in relation to local/regional contexts and capacities. This requires, on the one hand, paying attention to the agency of local actors, whilst also examining the broader economic and political relations – both historical and contemporary – which differentially locate rural places within wider global networks and power geometries (Massey 2005). It is to these historical relations which we briefly turn first and which suggest how, despite the contemporary focus of many accounts of globalization, Ceredigion has been embroiled in global movements of people and goods for centuries.

Ceredigion translates literally as 'land of Ceredig'; Ceredig being the son of Cunedda, a chieftain who regained control over much of northern and Mid Wales from Irish clans in the late fourth and early fifth centuries. Ceredigion was ruled by descendents of Ceredig until the early tenth century, when it was subsumed into the kingdom of Deheubarth. Following the conquest by Edward I

1 This chapter follows the OECD definition of rural areas as having a population density of fewer than 150 residents per square kilometre. A total of nine unitary authorities in Wales fall under this category and are taken to represent 'rural' Wales. These include: the Isle of Anglesey, Gwynedd, Conwy, Denbighshire, Powys, Ceredigion, Pembrokeshire, Carmarthenshire and Monmouthshire.

of England, Ceredigion (now referred to in the Anglicised form of Cardiganshire) became one of 13 counties within the principality of Wales from 1284 onwards. Politically, then, Ceredigion has featured as a disputed and strategically important territory throughout the lengthy history of the British Isles. Equally, the economic development of this distinctly rural part of Mid Wales is not one indicative of isolation and marginalization, and must be set in national and international contexts. As might be expected, farming has traditionally served as the basis of the Ceredigion economy, with dairying and stock-rearing of particular importance due to the region's poor soils. Prior to the arrival of the railway in 1866, juvenile cattle, pigs, sheep, horses and geese were taken by drovers on foot along tracks through the mountains to markets in England; Northampton, Rugby and London being important destinations for this trade (Moore-Colyer 1976). In this way, Ceredigion has long been deeply entrenched in the national agricultural economy and has played a foremost role in livestock production. Aside from farming, Ceredigion also has a lengthy and significant association with mining due to the county's important reserves of lead, silver and zinc. Metal mining in Ceredigion dates back to the Early Bronze Age, remaining largely small scale until the late sixteenth century when the discovery of silver provided the incentive for considerable expansion (Claughton and Freeman 2007). By the early nineteenth century the value of lead had outstripped the value of silver, becoming the principal focus of mining activities in the area and attracting a significant influx of Cornish miners as well as workers from far afield, including a large contingent from Italy. As the value of lead plummeted from the 1870s onwards the majority of mines ceased production, and, while a number of mines geared their extraction to zinc, competition from Australia and South America ensured that these enterprises came to an end by the 1920s. With over 50 miles of coastline as its western border, the sea has also shaped and mediated Ceredigion's role in the global economy, with a strong maritime tradition built on the dual pillars of fishing and trade. Indeed, given the state of the roads leading in and out of Mid Wales, trade was more effectively conducted by sea with the first established shipping routes emerging during the late thirteenth century when boats carried materials and provisions to castles established under the rule of Edward I. It was not, however, until the mid-eighteenth century that sea trade was firmly established as a cornerstone of Ceredigion's economic fortunes, with coastal trade bringing in coal and lime destined to improve poor quality agricultural land. Major exports included ores for smelting in Bristol and Swansea to the south, and oak destined for tanning factories in Ireland (Morgan 2005). During this period, ship building and chandlery became major local employers, with the majority of Welsh sailing vessels being constructed in the county. Not all materials, however, could be sourced locally, and the most significant import was Canadian pine, transported by local shipping companies on the return leg of transatlantic crossings. For the outward journey, the cargo was often emigrants (ibid.). Indeed, while the early nineteenth century was a prosperous period for seafaring in Mid Wales, there was much hardship in rural Ceredigion during the depression which followed the Napoleonic Wars. Consequently, many sought a better life overseas, and thousands emigrated to America, Canada, Australia and Patagonia, where a Welsh diaspora was established and evolved. It is evident then, that Ceredigion has never been truly insular; only perhaps insulated, to an extent, from short-term political-economic fluctuations and trends due to its peripherality and the relative stability this affords. During its more recent history, Ceredigion, like much of rural Wales, was profoundly impacted by broader productivist shifts in British agricultural practices, which saw farming becomingly economically unviable for many individuals. As Walford (2009, p. 61) describes:

> The region was already becoming economically marginalized by the 1960s, and the changing structure of British agriculture over the closing decades of the 20th century meant that agriculturalists in the area were farming on the edge of economic viability. Those remaining have become increasingly dependent on financial support for their production and for maintaining public goods, such as countryside and traditional landscape features.

The Mid Wales region (present-day Ceredigion, Powys and southern Gwynedd) was recognized by the British Government's Beacham Commission in the 1960s as a region in crisis, with low living standards and limited employment opportunities leading to severe depopulation (see Cloke et al. 1997, Day 1989, Walford 2009). The top-down response focused on encouraging development through improved transportation and communication links, industrial growth poles and increased manufacturing exports (Day and Hedger 1990), which, coupled with a national trend of counterurbanization that saw increasing numbers of retirees attracted to the scenic environment of Mid Wales, was successful for a time at stimulating employment and reversing depopulation. Indeed, by the 1980s, Ceredigion was experiencing population growth surpassing many other rural regions of the UK; although the concentration of this growth and associated infrastructural improvements around key settlements such as Aberystwyth left behind pockets of depopulation and deprivation in more remote inland areas (Cloke et al. 1997).

Woods (2009) notes, however, how many of the social and economic assumptions underwriting this post-war regional development strategy for rural Wales have since come to be challenged by processes of globalization. Indeed, whilst, as demonstrated, Ceredigion has long been entangled in global networks and connections, the nature and intensity of these connections is changing as they increasingly become a part of everyday life for many rural residents, transforming the economy and society of rural Wales in dynamic and uncertain ways. Specifically, Woods suggests it is possible to identify three key challenges that globalization has presented for the economy of rural Wales. These are: first, 'the liberalization of global trade in food and other commodities', which has increased pressure on traditional industries such as agriculture; second, 'new economic sectors losing out to global competition', including textile manufacturing in the 1970s and, more recently, ICT businesses and call centres; and thirdly, competition for jobs from 'internationalized labour migration', particularly the arrival of migrant workers from Central and Eastern Europe following the expansion of the European Union in 2004 (Woods 2009, pp. 172–3).

In the case of the latter, and in contrast to previous immigrant communities in the UK, the 'A8' migrant workers (from Czech Republic, Estonia, Hungary, Latvia, Lithuania, Poland, Slovakia and Slovenia) were not concentrated in cities, but more dispersed throughout the country, with around a quarter of Worker Registration Scheme members in rural districts (Commission for Rural Communities 2007).[2] In rural Wales, seven out of ten A8 workers were from Poland and over 70 per cent were aged under 35; this has particular resonance in an area where there was little or no significant history of immigration from outside the UK, as one migrant support worker suggested:

> You've also got to understand that people were negative here about people coming in from Manchester or Liverpool, so to come through from Poland or anywhere farther away … is also a big thing.

Wales Rural Observatory research with members of the migrant worker community found that, although the majority tended to have very positive views of the local community in that they had been made to feel welcome and safe, one in four reported experiences of discrimination (Woods and Watkin 2008). Most commonly this was verbal abuse in the workplace, with the most numerous occupations for migrant workers in Ceredigion being in low-paid service jobs, including cleaner/domestic assistant (19 per cent), catering/ kitchen assistant (18 per cent) and factory process operative (19 per cent).

This example of the poor opportunities afforded some migrant workers in rural Wales, and their occasional reporting of abusive treatment, is an indication of the uglier side of globalization processes that have facilitated the provision of cheap labour for the service economy dominating rural places. On the other (public-facing) side of the coin, the more widely reported feeling of being made welcome is harnessed by a tourism industry offering those service (catering, leisure and entertainment) jobs to migrant workers, as part of a rural identity of community, warmth and hospitality sold to potential visitors. Indeed, tourism promoting the natural assets, cultures and heritage of local place represents one form of endogenous response to the economic development challenges Ceredigion faces. In particular, the region's rich environmental capital in the form of protected coastline, estuary and mountainous areas provide opportunities for low impact eco-

---

2 The UK government stipulated that workers from A8 countries register on the Worker Registration Scheme within a month of joining a new employer, providing an indicative measure of economic migration from the new Accession States (although it does not measure self-employed workers, and a number of people may not have registered). Figures from the UK Border and Immigration Agency report that 19,060 of Worker Registration Scheme registrations were approved in Wales between May 2004 and March 2007 (3 per cent of the UK total).

tourism development and leisure pursuits (walking, mountain biking, kayaking) as part of a broader regional strategy for sustainable development. This includes the potential cache derived from the recent designation of the Dyfi UNESCO Biosphere in 2009, and specifically 'to use the Biosphere "brand" to promote the quality of local agricultural & other products and tourism experiences' (Dyfi Biosphere 2007).

In addition, then, to those globalization processes identified by Woods (2009), global climate change – and the increasingly global consciousness surrounding it – presents challenges and perceived opportunities for regional development in rural areas such as Ceredigion across multiple scales, for example through grassroots initiatives in eco- and local food tourism, to regional/national policy drivers for multifunctionality through renewable energy production and agri-environment schemes. Kitchen and Marsden (2009), however, suggest how the potential role of rural areas as spaces that can, and are required to, contribute towards adapting and mitigating global climate change is frequently held in check by the perception of an eco-economic paradox. This holds ecological concerns as an obstacle to rural social and economic development:

> Rural Wales possesses significant areas of high environmental value in various forms, including mountains, rivers, coastline wind and tidal resources and attractive landscapes. Nevertheless, it is particularly in areas with the most attractive and rich landscapes and ecologies that the problems of low skills and low economic activity rates are at their most extreme. (Kitchin and Marsden 2009, pp. 282–3)

The need to address this paradox is increasingly being recognized within rural development strategies for Wales. Specifically this is seen in the linking of environmental value and economic added value in fostering an integrative eco-economy, which adds value to natural resources found in rural regions in a sustainable fashion. This emphasizes the endogenous 'recalibration' of micro-economic behaviours and practices, which added together 'can potentially realign production-consumption chains and capture local and regional value between rural and urban spaces' (ibid.). Thus while a key aim of the EU regional development programme for West Wales is the promotion of business opportunities in relation to future environmental challenges and opportunities, different economic imaginaries and values may also be enrolled by local actors as integral to rural sustainable development; as examples from Ceredigion illustrate.

## Local Strategies for 'Success'

As noted, certain local contexts have contributed towards nurturing a strong environmental and social justice ethos in Ceredigion generally and specifically around the university town of Aberystwyth. Here, parts of the local community are engaged in practices that involve the recasting of relations with the carbon-based economy and are based on 'alternative' modes of value and exchange such as, for example, the swapping and 'upcycling' of unwanted goods to reduce landfill. Similarly, a desire to shorten linkages between production and consumption is evident through the high demand for local, often organic, farm produce at farmers markets and specialist food shops in a number of Ceredigion's larger towns. In meeting this demand for value-added produce, traditional agricultural enterprises may 'deepen' their activities and integration with broader rural eco-economic development concerns (Kitchin and Marsden 2009).

Further 'alternative' practices include Aberystwyth's involvement in the 'Transition Towns' network, which is premised on promoting community-led responses to climate change and the building of 'resilient and sustainable communities' through sharing skills and knowledge (Mason and Whitehead 2011), whilst more formalized opportunities for 'regional learning' are offered through the internationally recognized Centre for Alternative Technology in nearby Machynlleth. The presence of both the university and the Welsh Government's Environment and Sustainable Development department in Aberystwyth have also given rise to a research and knowledge cluster around issues of climate change mitigation and adaptation. This includes research undertaken at the university's Institute of Biological, Environmental and Rural Sciences into high-yielding biorenewable crops that are suited to growing on the region's generally arable-poor soils. The practical application of this research has the potential to add value to the land-based economy; whether this is seen as a 'broadening' of traditional rural enterprises remains to be seen.

At the same time, other potential dimensions of eco-economic activity are more embedded in 'mainstream' economic models, yet involve the selling of a locally embedded 'eco' identity to national and global markets. Luxury food tourism in the county is very much sold around the use of locally sourced ingredients with low food miles and quality providence, with establishments gaining reputations for both their hotel and food services. An example here is the Ynyshir Hall hotel, which falls within the Dyfi UNESCO biosphere area and directly adjoins an RSPB (The Royal Society for the Protection of Birds) nature reserve, and whose Michelin-starred restaurant extols the virtues of local produce:

> There's a wealth of superb, fresh local ingredients available locally making dining at Ynyshir Hall simply divine. Cardigan Bay teams with succulent seafood with lobsters, crabs and shrimp landed at Aberystwyth and Borth. There is wild salmon in the local rivers, venison and wild game are in abundance, the pork is all organic, the bacon is all hand cut for breakfast, sausages are all made by the local butcher to Ynyshir's own recipes. (source: www.ynyshirhall.co.uk)

Environmental concerns have also been factored into the business ethics and brand identity of several successful Ceredigion-based companies, namely *howies* clothing company and *Rachel's* organic dairy. For both of these, their embeddedness in place has been an integral part of the company's development and continued marketing strategy in which they respectively sell an image and associated lifestyle of 'rural Wales' to national and, increasingly, international markets. *Rachel's* is a producer of organic yoghurt and other dairy products founded on a dairy farm in Ceredigion, which sells itself on organic principles, traditional recipes and a harmony with nature found in rural/peripheral west Wales; the company's founding narrative recalls how yoghurt production began when snow storms prevented milk tankers reaching the farm to collect the milk, meaning it had to be used in another way (source: www.rachelsorganic.co.uk/about-us/our-story). *Rachel's* products are now stocked in most major supermarkets across the UK and despite the company being sold to American company Horizon Organic Dairy in 1999, and then to the French Groupe Lactalis in 2010 for a reputed asking price of £20 million, the majority of production continues in Ceredigion using milk supplied from local family farms.

*howies* active clothing company was founded by a group of outdoor enthusiast friends, with the intention to produce a high-quality product that would last longer and have a lower environmental impact by principally utilizing organic cottons. Based in the town of Cardigan, in south Ceredigion, its location next to the rugged coastline of Cardigan Bay is central to *howies* brand identity; reproduced through website and catalogue images of active twentysomethings playing in this landscape, such that the company is not only associated with a specific place but also an accompanying lifestyle for consumers to buy into. In 2006, following a period of rapid expansion, *howies* was invested in by the multinational clothing company Timberland in a deal that, crucially, allowed them to retain the company headquarters in Cardigan whilst now having the capital to expand into wider national and international markets. This investment also allowed *howies* to introduce previously unaffordable manufacturing standards such as environmental audits of their suppliers, which are now based in Portugal, China, Indonesia, Fiji and the UK, respectively. The subsequent sale of Timberland to the global brand conglomerate VF, however, led to the decision by *howies* original owners to buy back the company in January 2012 and continue operating from Cardigan as an independent company; becoming 'small again' in order to better pursue their commitment to sustainable business practices (this decision is detailed in a blog post 'On our own two feet', available at: www.howies.co.uk/brainfood/?p=22675).

The examples of *howies* and *Rachel's*, both (for a time) multi-nationally owned enterprises and both established in the rural economy of Mid Wales, are indicative of globalization as a force working through the countryside and not 'upon it'. Furthermore, the business development strategies of these companies have been very much centred on projecting rurality – and the Mid Wales wilderness more specifically – into the wider world through selected involvement in global supply chains. Selling their product and, particularly in the case of *howies*, an ideology, the manner in which these brands have prospered and grown is intrinsically shaped by the material conditions of their locality, and by the variegated cultures of the local economy. On the one hand, they are clearly competitive, specialized, export-orientated and integrated in the agglomerative tendencies of capitalism. On the other hand they are place-attached, culturally distinctive and environmentally conscious. The recent decision by *howies* to 'stand on their own' rather than becoming part of a global concern suggests, however, that these dual facets are not necessarily always compatible within a business ethos where social and environmental goals are factored alongside economic ones. In this way, neither company fits easily into a mainstream and arguably dominant economic narrative of a faceless, virulent neoliberalism, and nor do they adhere rigidly to those principles of exchange associated with 'alternative economic spaces' (Leyshon and Lee 2003) and so-called 'community economies' (Gibson-Graham 1996, 2006). Rather, they embody elements of both of these economic imaginaries (see Figure 5.1), and form part of the chaotic and amorphous amalgam of activities which constitute the 'actually existing economy' (Jessop 2004, p. 7), and which correspond to material interdependencies and the relationships between economic and extra-economic activities.

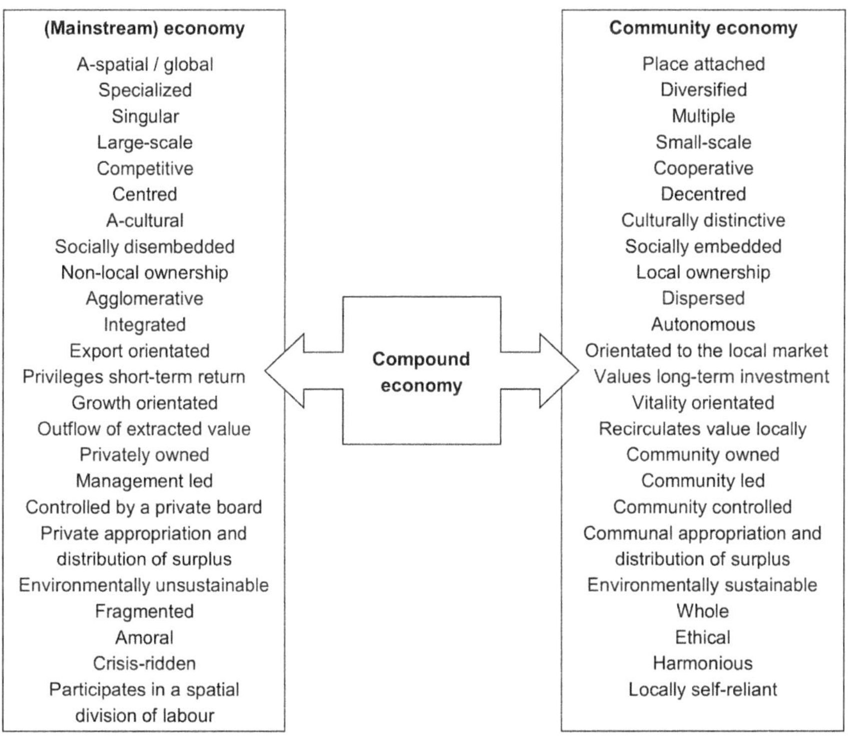


**Figure 5.1 Key words of economy and community economy**
*Source*: Heley et al. (2012, p. 371) after Gibson-Graham (2006, p. 87)

It is in this 'composite' economic space that forms of rural eco-economic development are seemingly being fostered in Ceredigion, where a range of environmental, cultural, social *and* economic values and aspirations come together through the micro-economic practices of various local enterprises, institutions and actors. Here, different endogenous strategies have been adopted for responding to the pressures of globalization and, at the same time, working within a context where spatial constraints of accessibility and connectivity, including limited broadband access in some areas, are continued factors in achieving 'successful' rural development. That is, development as variously defined in terms of growth, expansion, sustainability, security and/or resilience. There is more limited recognition of this composite space, however, in regional policy discourses, which encourages endogenous development yet are perhaps less sensitive to the particularities of local context in how this is made manifest.

Clearly, there are certain identifiable patterns of material interdependencies and economic cultures which allow us to designate regions and localities as being 'rural', and it is also the case that such discursive understandings play a role in governing what sorts of economic relations are forged and pursued. This process has political ramifications, and serves to shape the development trajectories of the countryside; particularly when these sets of assumptions are used as the basis for state-led interventions and regeneration strategies. In the case of Wales, Welsh Government (WG) policy has recently set out its understanding of the regions' economic character in the *Wales Spatial Plan* (2004; updated 2008). This document outlines the Welsh Government's objective to promote 'economic ambition' (WG 2008, p. 59) in rural Wales, the need for which is deemed as vital for the sustainable future of Mid Wales as a region characterized as having:

- A low 'critical mass' of population, which renders the region unattractive to inward investment;
- A high proportion of self-employment, which is identified as a hindrance to significant business growth;
- Poor external transport links and an under-developed ICT infrastructure, which is associated with marginalization from wider economic markets.

This account of Mid Wales' rural economy is clearly one which is aligned with mainstream economic understanding, foregrounding as it does a requirement for agglomerative, integrated and growth-orientated strategies. As a prompt for change and 'doing the rural economy' differently, however, it does not chime uniformly with the understandings, experiences and practises of those living and working in this region. This contrast is demonstrated in a recent survey undertaken in Mid Wales by WISERD (Welsh Institute of Social and Economic Research, Data and Methods), where attitudes among business owners towards their activities, aspirations and role in the local, regional, national and international economies was often in marked difference to the expressed outlook of the Wales Spatial Plan (Heley et al. 2012).

Among larger, established mass-production enterprises, stakeholders often identified distance from distribution centres and associatively high transport costs as being a barrier to competiveness in their market sector, and as factors which reduced their ability to attract inward investment. Prompting a shift into the production of premium goods as a means of compensating for relative inaccessibility, these same stakeholders also highlighted the possibilities afforded to companies in rural locations to successfully market and sell premium goods:

> There is no way we could manufacture a cheap and cheerful product on this site. Our product is premium, which warrants a premium price and can support slightly higher distribution costs. [Stakeholder: Food manufacturer]

Where representatives of larger companies underscored the role of premium goods in facilitating growth, small business owners tended to accentuate stability. Emphasizing long-term survival and a desire to achieve a good, but not exorbitant income, this group of respondents premised their continued fortunes on knowing what the community wanted and providing a good service, thus ensuring return trade. Recognizing that there was a small threshold population to support an enterprise, distance from other potential providers was identified as providing a stable market and security – even during periods of national and international recession:

> The downside of being a business in Mid Wales is that we don't see the extreme highs that the cities generate. On the plus side, come the downturn – like we are experiencing now, the market is generally much more stable. We don't get the highs and lows; the business stays on a much more even keel … [Stakeholder: Building supplies firm]

The 'protection' afforded by choosing to live and work in a peripheral location works alongside or in conjunction with other 'alternative' business ideals, with many traders stressing the social embeddedness of both their businesses and themselves in the local community alongside the difficulties and dangers of expanding into wider economic networks. Whilst business growth is therefore desirable, it was seldom discussed without reference to more overriding goals of survival and security:

> We are in business, and growth would be nice; it is something we don't look to stop and we are always looking to increase our business. You have to. But to fixate on growth and take your eye off the ball in terms of what you have already got is a disaster … There are four of us here who work full time, and all four of us own our own tiny little homes. We are in a business that looks to remain secure so that we don't have to worry about whatever mortgages we have. It is not about getting a new Bentley in 5 years time. We have good employment in a business we love, and as long as the business continues with a little bit of growth

> to cover inflation, rises in living costs and utility bills, we won't complain. Expansion is not the goal, security is. [Stakeholder: Music shop]

These more modest ambitions, linked to an appreciation of the positive affordances of a rural lifestyle despite comparatively lower income levels than other regions, were seen as making for more resilient ways of doing business in the current global financial context. Whilst this emphasis on mid- to long-term survival was often premised on a close-knit working relationship with the local population, local traders felt that national chains could not lay claim to such relationships and the reciprocal allegiance of local customers, and this was identified as a factor in their lack of resilience on the high street. As one shop owner put it:

> We know exactly what is going out of the door and how much is coming in, whereas bigger companies like Woolworths [recently defunct high street chain], they're just interested in turnover.

## Conclusions

Ceredigion and Mid Wales more broadly then, or rather the Mid Wales economy, are far from being globally marginalized or 'failing', either now or during its history. However, it is clear that it cannot and will not be able to compete on a production-orientated (post)-Fordist, mass-manufacturing playing field due to its structural locatedness within particular geometries of power. This is clearly articulated by stakeholders in this region, and is evident in their daily economic exchanges. As such, different endogenous strategies have been adopted by local actors for mediating the opportunities and threats presented by globalization processes, including the increasing global consciousness surrounding anthropogenic climate change mitigation. In some cases, the stability of an established customer base, knowledge of and responsiveness to customer needs, combined with the provision of more locally sourced (low-carbon) goods and services, is valued for providing potentially greater resilience to social, economic and environmental change. At the same time, other regional enterprises and entrepreneurs have been successful in building on and selling their local embeddedness, and, with it, an environmental consciousness integral to their business practices, as a way of successfully connecting into, and competing in, national and international networks of production and consumption.

Consequently we have seen here that the system of exchange is neither 'mainstream' or 'alternative', but rather a mixture of both; a situation that would arguably be the case in the majority of Europe's regions, be they rural or urban as we might understand them. What is also apparent is that local knowledge of people, practises and cultures are at play, and these are fundamental to successfully adapting to, and being resilient of, shifts in global markets and economic restructuring. It is also clear that this knowledge as employed by people living and working in Mid Wales is not necessarily in response to these shifts, but an integral part of them. Endogenous development strategies are thus not only about 'selling' local distinctiveness, but also the situated practice of this distinctiveness by local actors in terms of determining the shape and trajectory which development takes, and how its 'success' is to be variously defined.

## References


Barca, F. (2009) *An Agenda for a Reformed Cohesion Policy, A place-based approach to meeting European Union challenges and expectations.* Commissioner for Regional Policy, April 2009. Available at: http://www.europarl.europa.eu/meetdocs/2009_2014/documents/regi/dv/barca_report_/barca_report_en.pdf [accessed 14 January 2013].

Cheshire, L. (2006) *Governing Rural Development.* UK: Ashgate.

Claughton, P. and Freeman, M. (2007) *Ceredigion's Mining Heritage.* Available at: http://news.bbc.co.uk/local/midwales/hi/people_and_places/history/newsid_8306000/8306501.stm [accessed 14 January 2013].

Cloke, P., Goodwin, M. and Milbourne, P. (1997) *Rural Wales: Community and Marginalization.* Cardiff: University of Wales Press.

Commission for Rural Communities (2007) *A8 Migrant Workers in Rural Areas.* Cheltenham: CRC.

Day, G. and Hedger, M. (1990) 'Mid Wales: missing the point', *Urban Studies*, 27(2), 283–90.

Day, G., Rees, G. and Murdoch, J. (1989) 'Social change, rural localities and the state: the restructuring of rural Wales', *Journal of Rural Studies*, 5(3), 227–44.

Dyfi Biosphere (2007) *Dyfi Biosphere aims and objectives.* Available at: http://www.biosfferdyfi.org.uk/start/aims/ [accessed 14 January 2013].

European Commission (2009) *European Cohesion Policy in the United Kingdom.* Available at: http://ec.europa.eu/regional_policy/sources/docgener/informat/country2009/uk_en.pdf [accessed 14 January 2013].

European Commission (2011) *Operational Programme 'West Wales and the Valleys (Revised version).* Available at: http://ec.europa.eu/regional_policy/country/prordn/details_new.cfm?gv_PAY=UK&gv_reg=ALL&gv_PGM=1239&gv_defL=7&LAN=7 [accessed 14 January 2013].

Gibson-Graham, J.K. (1996) *The End of Capitalism (As We Knew It): A Feminist Critique of Political Economy*. Oxford: Blackwell.

Gibson-Graham, J.K. (2006*) A Postcapitalist Politics.* Minneapolis, MN: University of Minnesota Press.

Halfacree, K. (2007) 'Trial by space for a "radical rural": Introducing alternative localities, representations and lives', *Journal of Rural Studies*, 23, 125–41.

Heley, J., Gardner, G. and Watkin, S. (2012) 'Cultures of local economy in a Celtic fringe region', *European Urban and Regional Studies*, 19, 366–82.

Jessop, B. (2004) 'Critical Semiotic Analysis and Cultural Political Economy', *Critical Discourse Studies*, 1(1), 1–16.

Leyshon, A. and Lee, R. (2003) Alternative Economic Geographies, in Leyshon, A., Lee, R. and Williams C. (eds) *Alternative Economic Spaces*, pp. 1–27. London: Sage.

Kitchen, L. and Marsden, T. (2009) 'Creating Sustainable Rural Development through Stimulating the Eco-Economy: Beyond the Eco-economic Paradox', *European Journal of Rural Sociology*, 49(3), 273–94.

Marsden, T., Murdoch, J., Lowe, P., Munton, R. and Flynn, A. (1993) *Constructing the Countryside*. London: UCL Press.

Marsden, T. (1998) 'New Rural territories: regulating the differentiated rural spaces', *Journal of Rural Studies*, 14, 107–19.

Mason, K. and Whitehead, M. (2011) 'Transition urbanism and the contested politics of ethical place making', *Antipode*, 44, 493–516.

Massey, D. (2005) *For Space*. London: Sage.

Milbourne, P. (ed.) (2012) *Rural Wales in the Twenty-First Century: Society, Economy and Environment*. Cardiff: University of Wales Press.

Moore-Colyer, R. J. (1976) *The Welsh Cattle Drovers*. Cardiff: University of Wales Press.

Morgan, G. (2005) *Ceredigion: A Wealth of History*. Llandysul, Wales: Gomer Press.

Murdoch, J. (2006) Networking rurality: emergent complexity in the countryside, in Cloke, P., Marsden, T. and Mooney, P.H. (eds) *Handbook of Rural Studies*. London: Sage.

Ploeg, J.D. van der, Long, A. and Banks, J. (eds) (2002) *Living Countrysides: Rural development processes in Europe: the state of the art*. Amsterdam: Elsevier.

Walford, N. (2009) Connecting Rural and Urban Places: Enduring Migration between small areas in England and Wales, in Halseth, G., Markey, S. and Bruce, D. (eds) *The Next Rural Economies: Constructing rural place in a global economy*, pp. 59–74. CAB International.

Welsh Government (2004) *People, Places, Futures: The Wales Spatial Plan*. Cardiff: Welsh Assembly Government.

Welsh Government (2008) *People, Places, Futures – The Wales Spatial Plan 2008 Update*. Cardiff: Welsh Assembly Government.

Welsh Government (2011) *Local Area Summary Statistics: Ceredigion*. Available at: http://wales.gov.uk/topics/statistics/publications/localarea/?lang=en [accessed 14 January 2013].

Woods, M. (2007) 'Engaging the global countryside: globalization, hybridity and the reconstitution of rural place', *Progress in Human Geography*, 31, 485–507.

Woods, M. (2009) The political economies of place in the emergent global countryside: stories from rural Wales, in Halseth, G., Markey, S. and Bruce, D. (eds) *The Next Rural Economies: Constructing rural place in a global economy*, pp. 166–78. CAB International.

Woods, M. and Watkin S.R. (2008) *Central and Eastern European Migrant Workers in Rural Wales*. Cardiff: Welsh Assembly Government.

Chapter 6

# Constituting Connections and Rural Revitalization in 'The Comarca de Verín', Galicia (Spain)

María D. Domínguez García, Paul Swagemakers, Johannes S.C. Wiskerke and Xavier Simón Fernández

## Introduction

This chapter assesses the performance of the yet fragile compilation of connections between endogenous knowledge, natural resource management and new business models. This compilation has been constituted in the Comarca de Verín (Galicia, Spain) by applying a relational approach to understanding the reconstitution of rural places under globalization (Woods 2013). Rural development dynamics most often result from the mutually reinforcing processes of institutional and formal support and practitioners operating in informal networks (Giddens 1984). Therefore, this chapter examines connections between the pillars of regional learning (the public administrative sector, the knowledge support structure and stakeholders in the region (so-called Grassroots Development Initiatives) (GDIs) (Wellbrock et al. 2012)). In particular the focus here is on local initiatives that seek to support food and forest production. These productions are grounded in the preservation of one of Europe's traditional landscapes containing 'the complex history of a place or region, which still can be read from its composition and structure' (Antrop 2005, p. 25). In order to map the territorial characteristics of newly evolving connections, interface analysis is applied. The 'critical point of intersection between different lifeworlds, social fields or levels of social organization, where social discontinuities based upon discrepancies in values, interests, knowledges and power, are most likely to be located' (Long 2001, p. 243) is examined. An interface, in this framing, can take shape in a formalized, organizational entity or have an informal, often veiled appearance (Long 1999). In order to unravel the regional development dynamics, this chapter begins with a description of the case study area, followed by the discussion of the application of policy schemes. The constitution of the learning region is then assessed by analysing and discussing the connections between the three pillars of regional learning in the Comarca de Verín. The final section provides conclusions and policy recommendations on how to further develop an interpretative model of rural revitalization.

## Description Of, and Dynamics In, the Case Study Area

The Comarca de Verín (see Figure 6.1) is situated in the province of Ourense, in the southeastern part of Galicia, one of the 17 autonomous regions in Spain. A Comarca is a 'territorial intermediate environment formed by a set of neighboring municipalities with an internal cohesion among them based on geographical, historical, economic and functional facts' (Law 7/1996, 10th June). The Xunta de Galicia (the regional governance level) defines a Comarca as a territorial group of municipalities that represent a common history and form a practical entity regarding organization and planning; for the Comarca de Verín the eight municipalities are marked in Figure 6.1.

### *Crossroads*

For centuries, the Comarca de Verín has been at a crossroads and resting place for travellers between Castilla y Leon (northern Spain, Madrid) and Portugal (Porto, Lisbon). Nowadays, traffic consists largely of commercial haulage from Vigo (an important fish harbour and car factory in Europe), migrants driving 'home' (to Portugal) and tourists travelling to the tourist area of Rias Baixas (the coastal area in southwestern Galicia). Next to traffic passing Verín, the highway between Benavente and Vigo has improved the connection of Verín to Ourense (the capital city of the province (at 69 km)), Vigo (largest city in Galicia (at 166 km)) and to Santiago de Compostela (capital city of Galicia (at 173 km)). The most important

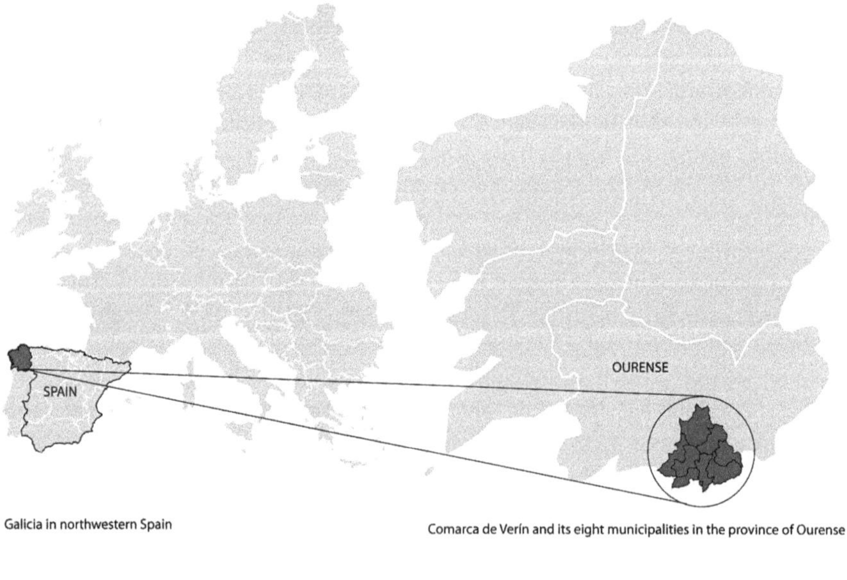


**Figure 6.1 Location of the Comarca de Verin**

airports in the region are in Vigo, Santiago de Compostela and Porto (175 km), which can all be easily reached by highway.

*Verín and its Remote Rural Areas*

The largely mountainous Comarca de Verín (1,007 km²) is bisected by the Monterrei Valley in which the town (Verín) is situated (Rolán Fernández 1994). Half of the population lives in this town and the rest in small villages (belonging to the municipalities Castrelo do Val, Cualedro, Laza, Monterrei, Oímbra, Riós and Vilardevós) scattered all over the Comarca. In the Comarca there are 175 officially recognized centres of population or singular entities (these are any habitable area within the municipal area, inhabited or not, clearly differentiated within the area) (INE 2010a). A point of major visual interest in the town of Verín is the Monterrei Castle from where the whole valley and all mountains can be seen. In the valley (with fertile soils which are frequently used for a diverse cropping of vegetables, but mostly potatoes and forage) many small rivers join the Támega, an alluvial complex of wetlands rich in flora and fauna, that belongs to Red Natura 2000. The mountains, characterized by peri-glacial shapes, have peaks covered by different species of heather. Its slopes have rich reservoirs of chestnut trees and pines. In particular this mountainous area is often abandoned (Corbelle-Rico et al. 2012) and consequently has an increased chance of fire.

*Socio-economic Dynamics in the Comarca*

In the past century different migratory waves have affected this region. In the beginning of the twentieth century out-migration to Latin America dominated. From the 1960s out-migration continued to European (Switzerland, France and Germany) and Spanish industrial areas (for example Barcelona, Bilbao, Madrid). Between 1960 and 2010, the population in the Comarca de Verín decreased from 43,285 to 28,596 inhabitants (INE 2010a, INE various years a). A lack of jobs and low economic activity rates characterize the socio-economic conditions in the Comarca. In 2001, the Comarca's average activity rate (39.6 per cent) was lower than the provincial (44.9 per cent) and regional rates (50.7 per cent), except in the municipality of Verín (48 per cent) (INE various years b, IGE 2005). These dynamics suggest that the seven 'satellite' municipalities suffer from severe socio-economic conditions. The population in the most remote municipalities declined by an average rate of over 3 per cent from 1996 to 2010 (excluding Verín) (INE various years a). Over the past 15 years, natural growth has been negative, due to a continuous negative trend of the number of births (IGE 2010, IGE 2010, INE various years a). In the same period, aging of the population increased. In 2010, over 30 per cent of the population was aged over 64 (INE 2010b). Depopulation, negative natural growth and the process of ageing impact on the maintenance of private and public (health) services in most municipalities. These negative socio-economic dynamics in the Comarca are confirmed by the *Tesorería General de la Seguridad Social* (the National Social Security System) that indicates an unemployment rate of 14.6 per cent (twice the Spanish average) in April 2008 (IGE 2010). The largest share of employment in the Comarca is in the service sector, especially in the municipality of Verín. In the rural satellite municipalities, agriculture still accounts for 22.9 per cent of employment (Cámara de Ourense et al. 2008).

## The Application of Policy Schemes in the Comarca de Verín

Public administration at the Comarcal level, including rural development policies, is geared to local, provincial and regional as well as national governance structure and supporting policies. Following Spanish accession to the European Union (EU) in 1986, a wide range of European, national and regional policy programmes aimed at revitalizing marginal rural areas and addressing negative demographic and socio-economic trends and environmental degradation have been implemented in Galicia. As a marginal area, the Comarca de Verín has been a beneficiary of these programmes. Figure 6.2 summarizes how this public support is mediated to beneficiaries in the Comarca: it displays the interactions between supporting policies, Grassroots Development Initiatives (GDIs) and the facilitation of joint learning and innovation (see arrows a, b and c).

For planning and rural development purposes, the Comarca is subordinated to various public governance levels: the *Concello* (the municipal level), the *Diputación provincial de Ourense* (provincial level), the *Xunta de Galicia* (regional level), the *Estado Español* (the central Spanish government, national level) and the EU (international level); and receives funding from both rural development and regional development programmes. In Spain, the funding of rural development includes horizontally implemented programmes covering the whole country, such as the *Fondo Español de Garantía Agraria* (the Spanish Platform of Agricultural Guarantee, which applies CAP regulations) and the *Dirección General de Desarrollo Sostenible del Medio Rural* (Directorate for Sustainable Rural Development, which coordinates regional authorities and regulations on co-financing programmes); and rural development programmes (RDPs) that are vertically implemented through the 17 autonomous regions. Support for regional development is targeted through the European Structural Funds to lagging regions. As a *convergence region*, Galicia benefits from the European Fund for Regional Development (ERDF), the European Social Fund (ESF) and the Cohesion Fund. The overall objective of the Galician RDP, derived from the Community Strategic Guidelines (CSG) and the National Strategy Plan, is to resolve the imbalance between the coastal areas where economic growth is concentrated and inland rural and mountain areas, which experience economic decline and depopulation. Its overall aim is to achieve a sustainable, strong and viable rural environment by 2013.

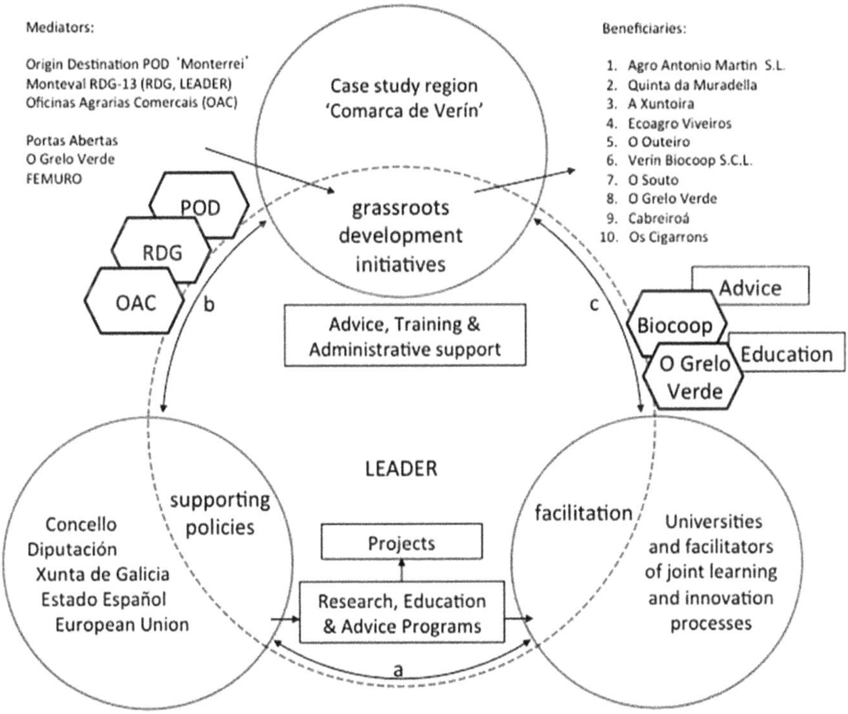


**Figure 6. 2 Interface for regional learning in the Comarca de Verin**

In the next section we further discuss the implementation of RDPs related to regional learning and the main 'public' stakeholders involved. We also develop an overview of the interfaces between public administration, the knowledge support structure and the development initiatives at the level of the Comarca (see also Roep and Wellbrock 2011, Wellbrock et al. 2010; 2011).

*Implementation Procedures of the Rural Development Programmes (RDPs)*

Rural development in Galicia is implemented through three grassroots programmes, funded by the EU, national and regional sources. The EU-funded LEADER programme is part of the RDP delivery across Galicia and is often applied together with nationally-funded PRODER and PRODER II programmes (*Programa de Desarrollo y Diversificación de zonas Rurales I* and *II*) that support endogenous sustainable development in rural areas (Objective 1 regions). The main objectives include the promotion of endogenous sustainable development in rural areas, strengthening and diversification of rural economy, population maintenance, income generation, social welfare and environmental conservation initiatives. Next, also a regional programme promoted by AGADER (*Axencia*

*Galega de Desenvolvemento Rural*, Galician Agency of Rural Development) has been introduced. This programme aims at supporting sustainable rural development in areas without LEADER groups. AGADER offices provide information and technicians in order to facilitate knowledge transfer and access to European, national and regional funds. The AGADER projects are similar to PRODER and LEADER projects but are fully financed by the Galician government. The AGADER programme aims at social participation, improvement of living standards, and promotion of human resources, environment and improvement of local economic activity.

In the Comarca de Verín the LEADER office houses *Asociación Monteval RDG-13* (Rural Development Group (RDG), named Local Action Group in earlier LEADER programmes), which applies the RDP (aiming at endogenous, sustainable and integrated development) and as such is responsible for implementing Verín's RDP in the period 2007–13. The RDG is the hub for local organizations and public and private stakeholders to gain public support under LEADER and similar policy schemes.

### *The Application of LEADER Programmes and the Constitution of Connections*

Since 1991, LEADER programmes have operated in the Comarca de Verín (LEADER I, II, LEADER+ and the current 2007–13 programme). In the 1990s, when the first LEADER programme was implemented, the NGO *Portas Abertas* played an important role. With the aim of facilitating rural development, *Portas Abertas* catalysed the start of many (often collective) GDIs; among them *Biocoop*, *A Xuntoira*, *Quinta da Muradella*, *O Souto*, *Ecoagro*, *FEMURO* (a platform of 70 women associations) and *O Grelo Verde*. The latter took an important role in disseminating knowledge on organic farming. Since LEADER+, however, the application of the LEADER programmes became a highly politicized issue and the initial organizational structure was ruptured. From then onwards the *Asociación Monteval RDG-13* implemented the programmes and organized support for GDIs in the Comarca. Hence, *Portas Abertas* and *O Grelo Verde* lost their leading positions with a consequent reduction in access to European funding. Many of the people involved feel that their efforts on improving the rural economy have become increasingly neglected. These 'early' GDIs still exist but without receiving LEADER support. Nowadays, they derive knowledge in various, often informally organized networks. The Universities of Santiago de Compostella and Vigo for example participate in projects on natural resource management (the chestnut production) and socio-economic dynamics (avoiding the negative impact of 'abandonment' of remote rural areas). These efforts are, however, poorly recognized by local stakeholders. As the case study research reveals, the villagers tend to 'hang their ears' to 'the ones' that manage the money coming from structural, cohesion and CAP funds.

Over the past 20 years, the application of policy programmes has resulted in new socially constructed interfaces among stakeholders and governmental bodies.

Simultaneously these programmes have catalysed the constitution of connections between public administration, (in)formal knowledge structures and GDIs.

## Develop Potential in the Comarca de Verín

In this section we discuss a series of Grassroots Development Initiatives (GDIs) with respect to the preservation of natural resources and contributions to social cohesion and engagement in the 'global economy'. By applying a 'snowball approach' we were able to identify a wide range of initiatives in the Comarca, including relatively long-established groups and more recent start-ups. Information about these initiatives was collected through key-informant interviews and documentary analysis. In total 20 respondents were interviewed: in addition to the ten selected grassroots initiatives in this paper interviews were conducted with representatives of the PDO Monterrei (wine), the comarcal agrarian office (former extension services), a professor at the University of Vigo specialized in chestnut tree plantations, the Local Action Group of the LEADER programme, four board members of the NGO Portas Abertas, which supported O Outeiro, Quinta da Muradella and O Souto in the first LEADER programmes implemented, and two school teachers. All the interviews were recorded and transcribed, after which a content analysis was applied on endogenous knowledge, natural resource management and the creation of business models.

The ten selected initiatives (listed as beneficiaries in Figure 6.2) are concerned with the natural and socio-cultural environment. Some, in particular, relate to *monte* (that is the higher situated areas most often consisting of forests and bushes) and the potential contribution to the local economy of the use of this often abandoned resource base. In-depth, semi-structured interviews were conducted with the representatives of each of the GDIs. These interviews are then used to form the basis of the discussion on globalization as a combination of processes, conditions and discourses or ideologies (each of them significant in influencing regional rural development trajectories).

### *The Ten Selected Initiatives*

1. For 30 years, *Agroalimentaria Antonio Martin S.L.* merchandizes quality food products, mainly Galician and northern Portuguese salted pork hams. During the chestnut season the owners buy chestnuts gathered by local pickers, selecting chestnuts on size and quality, and then exporting to France and Italy.
2. Some 20 years ago, in 1991, *Quinta da Muradella* was founded. This is a wine cellar which recovers traditional vineyards and endogenous varieties in order to produce quality wine that is sold in niche markets worldwide. The production of the grapes in the remote vineyards, based on endogenous knowledge attained from villagers, is processed in a modern wine cellar in

the town of Verín. The owner studied viticulture, enology and a Masters study in wine production. He combines this knowledge with the traditional knowledge that the villagers share with him, and provides labour to two fixed and five seasonal workers.

3. Also founded in 1991, the family enterprise *Xuntoira Mobiliario de Cociña S.L.* has developed into a wood factory with a cooperative structure. It is situated in the remote village of Berrande (municipality of Vilardevós) and provides labour to 60 workers who mainly live in Verín. The raw materials (wood) are imported. Activities have shifted from carpentry into the specialized production of kitchen doors.
4. In 1993, the cooperative plant nursery *Ecoagro-Viveiros* (at that time 10 people) started with LEADER funding producing parent material for vegetable gardens. It converted into a company in 2001.
5. Since the late 1990s, organic farm *O Outeiro* has shifted from dairy to beef production (mainly Vianesa breeds). Due to fields being small and scattered all over the village (a typical Galician characteristic), 100 hectares of land owned by several family members are in use to pasture about 100 autochthonous suckle cows. The bulls are fattened, slaughtered and sold as organic quality beef in short food supply chains. Subsidy programmes support these activities, that is, providing possibilities for Galician farmers to use autochthonous breeds.
6. In 1998, the beef production cooperative *Verín Biocoop S.C.L.* was founded. It gradually has converted into a major regional promoter of the marketing of organic meat (beef products are processed and sold mainly outside the Comarca), recovery and use of autochthonous breeds and the service and advice regarding the application of organic certification.
7. In 2001, *O Souto* was founded, in which 60 per cent of the chestnut tree owners in Vilardevós (the most mountainous village in the Comarca) are associated (with a total of 40,000 out of 100,000 trees in the village). The aim of *O Souto* is to improve the condition of the plantations, the classification of varieties and the revenues from individual fruit picking.
8. Since 2001, *O Grelo Verde* (an NGO fostering organic food production and consumption with 15 to 25 active members) provides and shares knowledge on organic farming in courses and advisory consultancy. This NGO seeks to sustain local food production and consumption by organizing organic school meals and listing organic producers and retailers (all over Spain).
9. In 2006, *Cabreiroá* (since 1906), a mineral water bottling plant, was bought by Hijos de Rivera Ltd. Using the new company's marketing channels, Verín's water nowadays is sold in a much wider geographical area. The company provides labour to 44 workers, almost all living in the Comarca.
10. The final initiative included the preservation of traditions related to Carnival (Verín is regionally known for its Carnival tradition that dates from the middle Ages): *Os Cigarrons*. It currently receives local, public and private support for organizing 'obradoiros': special workshops on the manufacture

and maintenance of locally specific suits and masks (about 500 suits in the Comarca de Verín) that are used in the annually parade in the town of Verín.

*The Constitution of Connections*

Over the last 20 years, many GDIs have started and matured. The connections between these initiatives, as well as connections between the initiatives and public support, have, however, been poorly developed. Consequently, employment, added value and multiplier effects failed to materialize. Although all ten initiatives contribute to regional economic activity, the landscape remains the region's 'undervalorised' resource base. Hence, it is important to further explore how the aforementioned connections can be strengthened and developed in relation to the reproduction of the traditional landscape.

In the Comarca, a notable initiative that combines sustainable natural resource management, social cooperation and a high economic performance is *Biocoop*. Over the past 15 years, this cooperative has attracted many farmers to join. It has developed from a regular farming cooperative (founded by the owner of the *O Outeiro* farm, a well-known initiator in Galicia) into a specialized provider of organic and endogenously oriented production techniques (Dominguez et al. 2012, Swagemakers et al. 2012). It has an office in the subsidized *Centro de desenvolvemento Cooperativo Parque Empresarial de Pazos* (the Regional Centre for Cooperation) in Verín. Its main activities are the implementation of organic certification among its members and the recovery of autochthonous breeds that complement wider efforts to recover the natural (and traditional) *monte* landscape. The *monte* area, mostly made up of trees, bushes and scrub (Bouhier 1979, Soto 2006), and often abandoned, represents a large underutilized resource base for Galician farm and forest production. The president of *A Xuntoira* argues, for example, that if the forest sector had been better organized, millions of euro could have been paid to local tree owners through the collection and collective marketing of chestnuts (as in the *O Souto* initiative) and even more so by producing and processing local wood. This could have resulted in an important multiplier effect for the local economy. The presidents of *O Souto* and *A Xuntoira* wanted to start plantations but until today this remains difficult: private and communal land-ownership limit the access to the land, which is very often unutilized. Governmental programmes for land consolidation (Coimbra 2011, Meijer 2010) have been proposed by *O Souto* and were approved after several years of discussion in the *Xunta de Galicia*. Further implementation of these plans however will however take many years (President of *O Souto*, personal communication). Social aspects seem to limit the successful implementation of rural development programmes. *O Souto* and *Os Cigarrons*, although different in character, provide some understanding of how the implementation of regional learning in the Comarca might work. The initiators of *Os Cigarrons* for example focus on how local cultural values and its material aspects (suits and masks) can be maintained. The experience on how people can be approached and involved in protecting socio-cultural values could perhaps be useful in setting up innovative programmes for the maintenance of the traditional landscape.

Initiators indicate that all progress starts with ideas about how a situation can possibly be improved, and meeting people with similar ideals and knowledge on how to improve and catalyse territorial rooted development:

> We started organizing people because of our concerns on the chestnut production and the potentials we saw, whilst the administration [regional government] did nothing. (President O Souto)

> When I started working at Portas Abertas, I discovered more people thinking like me. From then onwards I started to look for other locals who produced wine using traditional varieties – which at that time were rapidly disappearing. (owner Quinta da Muradella)

However, as knowledge on natural resource management is context-dependent (van Kessel 1990) and in essence incomplete and thus temporary in nature (Long 2001), its potential is difficult to predict. Whilst the outside world's ignorance about, or the underestimation of, such knowledge can be a source of frustration and aversion for those developing this knowledge (Wynne 1996), public support *can* help practitioners move forward, as often is the case with (EU) subsidies:

> I was a village councilor and nobody else wanted to carry out the project that had been approved by LEADER. (owner Ecoagro Viveiros)

> Subsidies are necessary to take a first step, to overcome indecision, to support. (owner Agroalimentarias Antonio Martín)

> AGACA [Galician association of cooperatives] was of big help giving us advice. (President Verín Biocoop)

The use and impact of subsidies differs however: a business can be started (as for *Ecoagro Viveiros*), the direction of its activities geared (as for *Agroalimentarias Antonio Martín*), or up-scaled (as for *Verín Biocoop*, *O Souto*, and *O Grelo Verde*). Until the 1940s, *Cabreiroá* (the water bottling company) for example has been attracting thousands of tourists to their hotel-spa. Nowadays, its activities are limited to selling drinking water and hence it has lost its wider economic impact on the Comarca. One of the most strongly and independently developing sectors in the Comarca is wine production, especially after the creation of the *Designation of Origin (POD) 'Monterrei'* in the 1990s. *Quinta de Muradella* goes beyond just wine production and the *POD 'Monterrei'*: by recovering traditional vineyards and ancient varieties it contributes to the preservation of *monte* and the traditional landscape in general. Additionally, its initiator has sought to harness endogenous knowledge among villagers who are no longer able to manage the vineyards themselves and combines this with selling the wine in worldwide niche-markets. Combining and integrating viticulture with other activities that can bring (otherwise abandoned) areas back into production could result in sustainable economic growth in the Comarca. Most of the marketing opportunities for businesses are situated outside the region. *O Grelo Verde* therefore has explored marketing opportunities for food products in the Comarca itself. With financial support from the regional government, *O Grelo Verde* ran a pilot project to supply primary schools with organic produced food. The programme ran successful for two years and won interest among other schools in the region until administrative and bureaucratic problems disrupted the scheme. Nonetheless, most of the initiators emphasize the crucial importance of public funding and the support of institutions such as the *Oficinas Agrarias Comarcales* (OAC), *POD 'Monterrei'*, *Monteval RDG-13* (and especially its predecessor, the LEADER, financed *Portas Abertas*), *Asociación Galega de Cooperativas Agrarias* (AGACA) and other, mostly European-funded initiatives. Without public support many of the GDIs would never have started.

## Building on Niche Sectors?

In this chapter it has been illustrated how GDIs in the Comarca de Verín operate in informal networks and aim to develop and sustain business activities that are based on natural resources by building links with local, regional and even global markets. Drawing on the broad findings of DERREG, Woods (2013) outlines a typology of regional engagement with the global economy that involves forms of collective action and territorial expression, such that they excel the actions of individual firms, organizations or households. Within this typology, the Comarca de Verín best fits the model of 're-localizers' (see also Chapter 12 book for a description and distinction of re-localization as development strategy). *Agroalimentaria Antonio Martin, Biocoop*, *O Outeiro* and *Quinta da Murella* are GDIs that have realized their own marketing channels. For example, *O Outeiro* uses autochthonous breeds which are sold by *Biocoop* all over Spain, and *Quinta da Muradella* uses and recovers 'old' wine varieties (and endogenous knowledge on the exploitation of traditional vineyards) to export a small production volume of selective wines all around the world. *O Grelo Verde* experimented with local (and organic) food procurement and would have liked to further organize the direct selling of organic food products under such schemes, but under the current funding regime had no opportunities to continue these activities. They continue promoting organic food products in general, and local food production in particular. They also train people to produce organically. *O Souto* aimed to foster the chestnut economy of the region, realizing added value to what we have defined as the traditional landscape by creating money flows directly to small-scaled producers through collective bargaining on price. Recently, their efforts were picked up by others and chestnuts can now be sold, individually, under the banner of 'Protected Geographical Indication'.

Besides the classification of re-localization, these dynamics represent Woods' (2013) classification of 'niche-innovation' and incorporate also elements of 'global conservation'. In niche-innovation rural regions engage with the global economy by producing and exporting goods to serve global niche markets, or attracting visitors to niche events. Global conservation refers to localities that have been integrated – willingly or unwillingly – into international structures for nature conservation and environmental protection, informed by global environmental discourses.

AGADER (*Xunta de Galicia*) supports the territorial promotion of the Comarca, including local tourist facilities (lodges, restaurants, routes) and typical food products (chestnuts, wine and meat products). The business models developed by the earliest niche-innovators perfectly fit the discourse of rural development programmes. Yet, despite their individual success, they remain relatively isolated from public administration and knowledge infrastructure. Their efforts to integrate nature protection, successful business models and social innovations remain poorly actioned. Whilst the incorporation into business strategies of autochthonous breeds, traditional vineyards and the recovery of the productivity of chestnut plantations and *monte* in the Comarca possibly would win interest of, and could be followed up by, rural dwellers, the earliest niche-innovators remain outside current rural development programmes in the region, as do their business orientations. The alternative model for rural development constituted by these early niche-innovators remains unarticulated. The mostly elderly people that live in or migrate back to the Comarca de Verín are either disinterested or remain unaware of the economic development potentials in the region. A lack of institutional and formal support worsens the situation: policymakers often conclude that rural dwellers hardly or do not opt at all to reorganize access to the natural resource base, and allow outsiders to implement highly water consumptive, non-autochthonous 'industrial' agro-forestry activities (Instituto Energético de Galicia 2012, p. 73). In the study area for example, the same report promotes *Eucalyptus nitens* and *Robinia pseudoacacia* plantations (ibid., pp. 76–7) with the argument that these plantations will reduce land abandonment and bring economic benefits to the region. These species however do not comply with the visions and the implemented rural development practice as initiated by, for example, *O Souto*. This is not an exceptional situation: in the initial, innovative phase of a system change, often mainstream knowledge structure and policies are disconnected from the hidden innovative practices. The breakthrough of such niche-innovations requires the stabilization of learning processes in a dominant design combined with powerful actors joining the support network (Geels and Schot 2007). In the conditions of Europe's financial crisis, young people might migrate back into rural areas and search for opportunities to build their future. In this situation, niche-actors, in our analysis categorized as 're-localizers' and as (potential) knowledge brokers, *might* play a decisive role in periods that follow the phase of niche constitution: scaling up their initiatives (with the support of young people migrating back to the countryside) and incorporating these in more formal knowledge infrastructure and in policy frames.

**Conclusion and Recommendations**

As in many rural regions like the Comarca de Verín, economic, social and environmental dynamics are exposed to globalization processes. The opportunities and threats resulting from these processes influence rural development policies and practices. By focusing on the endogenous development potential, we have analysed the (in)formal interfaces between the public administration, the knowledge infrastructure and Grassroots Development Initiatives (the three pillars of regional learning) and have discussed how these contribute to a model of rural revitalization based on re-localization of networks and resources, niche innovation and conservation.

In the Comarca de Verín, public support has played a decisive role in enabling a group of rural dwellers to become successful entrepreneurs. However, programmes implemented by the Galician government have become politicized. Local power-holders increasingly use funds to increase their popularity and their possibility of being re-elected for positions in the public domain while avoiding the incorporation of certain groups and/or individuals but assuring project benefits to others. Until now, formal research, education and advice programmes and projects remain limited in strengthening endogenous rural development potentials. Supportive policies implemented via the RDG, POD and OAC also have had a limited impact on strengthening the GDIs in relation to the maintenance and utilization of the traditional landscape. This has consequences for the assessment of the learning region by means of the formally created and visible relations between public administration, knowledge structures and grassroots initiatives. In the study area, an informal and to a large extent invisible network of actors operates outside the formalized sphere. The case study research clarifies how the innovative grassroots initiatives hold the potential to constitute connections between the three pillars of regional learning and can effectuate sustainable rural development if they would again be empowered with formal institutional support. Their activities make a difference regarding the conservation of the traditional landscape which, in the eyes of the initiators, holds the promise of bringing a common future. In addition, it would be useful to calculate the percentage of initiatives that contribute to the local economy and the extent to which national global firms have infiltrated the local economy. In such a performance assessment, alongside an economic impact assessment in terms of economics of scope, the social and ecological effects of business models should also be included. Further, in order to optimize the interpretative model of rural revitalization, research is needed to explore, test and verify interrelations between the adjustments made by Grassroots Development Initiatives in cases such as the Comarca de Verín, and, for example, in European Innovation Partnerships (EIPs) in the programming period 2014–16. Meanwhile, politicians and policymakers should pursue an objectives-led policy that allows for and stimulates the exchange of ideas, solutions and the creation of knowledge networks and promotes research on endogenous development potentials. When endogenous knowledge and local effort are sufficiently supported by the public administration and the knowledge support structure, a vital and sustainable rural constellation might further develop.

**Acknowledgements**

We are grateful to Joana Covelo Alonso (architect) for providing the map in Figure 6.1.

**References**

Cámara de Ourense, Concello de Verín and Antena Local de Verín (2008) *Estudio Socio-económico. Plan de Actuación Verín 2009–2014.* Cámara de Ourense, Concello de Verín and Antena Local de Verin. Available at: www.verin.es/fixo/ADL/plan2009–2014.pdf [accessed 3 December 2011].

Antrop, M. (2005) Why landscapes of the past are important for the future, *Landscape and Urban Planning*, 70, 21–34.

Bouhier A. (1979) *La Galice. Essay geographique d'annalyse et d'interpretation d'un vieux complex agraire*. La Roche-sur-Yon: Imprimerie Yonnaise.

Coimbra, E. (2011) Land management grounded in social soil. Galician land banking in context, *Sociologia, Problemas e Práticas*, 66, 135–55.

Corbelle-Rico, E., Crecente-Maseda, R. and Santé-Riveira, I. (2012) Multi-scale assessment and spatial modelling of agricultural land abandonment in a European peripheral region: Galicia (Spain), 1956–2004, *Land Use Policy*, 29, 493–501.

Domínguez García, M.D., Swagemakers, P., Bock, B.B. and Simón Fernández, X. (2012) Making a living: Grassroots development initiatives, natural resource management and institutional support in Galicia, Spain, *European Countryside*, 4(1), 17–30.

Geels, F.W. and J. Schot (2007) Typology of sociotechnical transition pathways, *Research Policy*, 36, 399–417.

Giddens, A. (1984) *The Constitution of Society: Outline of the theory of structuration*. Cambridge: Polity Press.

IGE (2010) *Movemento natural da poboación*. Instituto Galego de Estatística. Available at: www.ige.eu [accessed 5 October 2011].

IGE (2005) *Banco de datoscomarcal. Comarca Verin. Taxas actividade, ocupación e paro de poboación de 16 e mais anos*. Instituto Galego de Estatística. Available at: www.ige.eu [accessed 11 October 2012].

INE (2010a) *Nomenclator*. Instituto Nacional de Estadística. Available at: www.ine.es [accessed 4 October2011].

INE (2010b) *Padron Municipal 2010*. Instituto Nacional de Estadística. Available at: www.ine.es [accessed 4 October 2011].

INE (various years a) *Padrón Municipal*. Instituto Nacional de Estadística. Available at: www.ine.es [accessed 4 October 2011].

INE (various years b) *Movimiento natural de la poboación*. Instituto Nacional de Estadística. Available at: www.ine.es [accessed 5 October 2011].

Instituto Energético de Galicia (2012) *Análisis de la viabilidad del mercado de biomasa en Galicia y norte de Portugal*. Available at: http://www.inega.es/publicacions/enerxiasrenovabeis/publicacion_0029.html [accessed 28 November 2012].

Kessel, J. van (1990) Herwaarderen om te herleven. Productieritueel en technologisch betoog bij Andesvolkeren, *Derde Wereld*, 1–2, 77–97.

Long, N. (2001) *Development Sociology, Actor Perspectives*. London: Routledge.

Long, N. (1999) *The Multiple Optic of Interface Analysis*. UNESCO background paper on Interface Analysis. Available at: http://lanic.utexas.edu/project/etext/llilas/claspo/workingpapers/multipleoptic.pdf [accessed 3 December 2012].

Meijer, M. (2010) *Balancing Between Governing Styles. Participatory practices in rural Galicia*. MSc thesis. Wageningen: Wageningen University.

Roep, D. and Wellbrock, W. (2011) *Deliverable 4.4 Summary Report*. DERREG Developing Europe's Rural Regions in the Era of Globalisation. Available at: http://www.derreg.eu/system/files/D4.4%20Synthesis%20report.pdf [accessed 3 December 2012]

Rolán Fernández, A. (1994) *Verín y comarca*. León: Guia Everest. Editorial Evergráficas.

Swagemakers, P., Domínguez García, M.D., Simón Fernández, X. and Wiskerke, J.S.C. (2012) Unfolding farm practices: Working towards sustainable food production in the Netherlands and Spain, *Journal of Agriculture, Food Systems and Community Development*, 2(2), 129–45.

Wellbrock, W., Roep, D. and Wiskerke, J.S.C. (2012) An integrated perspective on rural regional learning, *European Countryside*, 4(1), 1–16.

Wellbrock, W., Roep, D., Kairyte, E., Farrell, M., Mahon, M., Frys, W., Nienaber, B., Kriszan, M., Nadler, R., Burdack, J. and Dominguez Garcia, M.D. (2010) *Deliverable 4.1 Overview of learning and innovation support strategies*. DERREG Developing Europe's Rural Regions in the Era of Globalisation. Available at: http://www.derreg.eu/system/files/DERREG%20D4%201%20Overview%20of%20learning%20and%20support%20strategies%20%20(25–06–2010)_0.pdf [accessed 3 December 2012].

Wellbrock, W., Roep, D., Kairyte, E., Farrell, M., Mahon, M., Frys, W., Nienaber, B., Kriszan, M., Nadler, R., Burdack, J. and Dominguez Garcia, M.D. (2011) *Deliverable 4.2 Support of Joint Learning and Innovation in Grassroots Development Initiatives: Operational quality of arrangements*. DERREG Developing Europe's Rural Regions in the Era of Globalisation. Available at: http://www.derreg.eu/system/files/D4.2%20Summary%20analysis%20of%20support%20to%20joint%20learning%20and%20innovation%20(final%20version%20June%209%202011)_0.pdf [accessed 3 December 2012].

Woods, M. (2013) Regions engaging globalization: a typology of regional responses in rural Europe, in Ramsey, D., Evans, N., Abrams, J. and Clark, J. (eds) *Understanding Change and Development: Perspectives from rural geography*. Brandon: Rural Development Institute.

Wynne, B. (1996) May the sheep safely graze? A reflexive view of the expert-lay knowledge divide, in Lash, S., Szerszynski, B. and Wynne, B. (eds) *Risk, Environment and Modernity*, pp. 44–83. London: Sage.

# Chapter 7
# Greater Globalization Challenges – Lesser Rural Responses: The Case of Alytus County, Lithuania

Emilija Kairytė

## Introduction

Alytus county, in the south-east of Lithuania, is a lagging region with relatively low incomes, high unemployment, falling population and net out-migration. It represents a region facing significant social and economic challenges from the impact of intensified globalization applied to a context of post-Soviet recovery in a period of global recession. As this chapter will discuss, the capacity of the region to respond to these challenges, and the implications for the region's socio-economic development, are strongly shaped by its geographical and historical setting as a post-Soviet economy located on the eastern border of the European Union.

Copus (2001) and others have argued that processes of global integration have undermined conventional core-periphery theories of economic development and promoted new models of polycentric development. Similarly, relational perspectives on globalization, discussed in Chapter 1 of this volume, have emphasized the continuing significance of local agency in influencing globalization outcomes. These theoretical arguments are important in suggesting that regions such as Alytus county that occupy locations on the geographical peripheries of nations and/or of the European Union should not be written off, but should have the potential to compete with more centrally located regions. Yet, contextual data shows Alytus county lagging economically and losing population through net out-migration. As such, the chapter explores the capacity of local agency in Alytus county and identifies the structural constraints that still limit its scope to act.

To pursue this aim, the chapter draws on research carried out by the Institute NeVork team during 2009–11 on two themes in the DERREG project: global engagement and local embeddedness of rural business and international mobility and migration of rural populations. As part of the first theme, research in Alytus county investigated the spatial and 'relational' structures of the business networks of small and medium-sized enterprises (SMEs) in rural areas through a questionnaire survey and interviews, exploring whether emerging rural business network configurations revealed any evidence of positive engagement in the global economy. For the second theme, research in Alytus county focused on migrant workers (including migrant workers returning from Western Europe) and established 'foreign' migrants from other parts of the former Soviet Union. The study explored the backgrounds and migratory pathways of these individuals, and also examined their social and economic integration in the region (see Chapter 1 for more details about these themes).

The overall objective here is to discuss how these interconnected challenges change and reposition the region in the 'global countryside' and to consider the factors determining regional abilities to respond to these challenges.

## The Context of Alytus County

Alytus county has international borders with Poland and Belarus. Although the county administration was abolished in a reorganization of local government in July 2010, the county still forms the basis of regional identity and is still used for statistical purposes. Its population in 2011 was 167,261, the majority of whom lived in the city of Alytus – an urban municipality with a population of 98,000 – the remaining residents being dispersed across four rural municipalities. The urban majority is a result of recent urbanization and the region is still currently categorized by the OECD as 'predominantly rural' (Eurostat 2010), having the third lowest population density of Lithuanian regions at 30.8 persons per $km^2$. The landscape of Alytus county is also characteristically rural. Nearly half of its territory is covered by forest and there are several large lakes, as well as sections of the Nemunas and Merkys Rivers. One-third of the land area is suitable for agriculture and it has the lowest share of agricultural production of all the regions of Lithuania (Statistics Lithuania 2011a). Farming includes grain, potato and other vegetable growing, as well as cattle for meat and dairy. Berry and mushroom picking in forests are important additional economic activities alongside wood processing, with farmers also diversifying into nursery gardens, cultivating medical and spice herbs, breeding fur animals and developing industrial lake fishing (Governor Administrator of Alytus 2011). Consequently, exportation of wood products and furniture, frozen berries and processed mushrooms are typical for the region.

The natural environment of Alytus county includes a number of national parks, regional parks and other protected areas, including a UNESCO biosphere reserve. The region is also known for its vernacular architecture, folk traditions and cultural heritage. These features make it attractive for tourism and outdoor sports activities, with notable sites including the Druskininkai thermal spa and all year Snow Arena, as well as Grūtas Park, with its historical exhibition of Soviet-era sculptures displayed in a forest setting.

The geographic location of the region is both a strength and weakness. The county is strategically positioned for transport connections between Lithuania and neighbouring states: an international cargo transhipment station at Šeštokai–Mockava connects Lithuania with the European railway network and the railway line from Belarus to Vilnius passes through Marcinkonys and Varėna; international border crossings link highways to Central and Western Europe, via Poland, and to the east, via Belarus; whilst the navigable Nemunas River connects to the seaport of Klaipeda. Overall, Alytus county can be classed as being relatively peripheral within Lithuania, although it is less than 100 kilometres from the major cities of Kaunas and Vilnius, with their international airports. Its neighbouring regions in Belarus and Poland are also sparsely populated and economically marginalized. Furthermore, as the European Union's eastern border, the frontier with Belarus has become ever more tightly controlled in terms of the movement of people and goods.

*Social and Population Characteristics and Trends*

While the population of Lithuania as a whole is decreasing, Alytus county recorded the second greatest fall in population of the ten counties in 2010, at 3.6 per cent – a reduction of 6,165 people, continuing a trend of depopulation that has persisted since 1991 (Statistics Lithuania 2011a). The population has shrunk by over 30,000 people in two decades. International net out-migration has contributed to this trend, with out-migrants out-numbering in-migrants by nearly 5,000 in 2010 (see Table 7.1). These include 5,153 emigrants who left Alytus county to move to another country in 2010, compared to just 242 direct immigrants into the county from other countries. A significant increase in the number of registered emigrants in 2010 reflects, in part, a new legal requirement introduced by the Law on Health Insurance for permanent residents of Lithuania to pay compulsory health insurance, encouraging Lithuanians working abroad to declare themselves to be non-resident.

Out-migration has also strongly contributed to the demographic aging of the population in Alytus County (Vagnorienė 2010). In January 2011, the ratio of elderly people (aged 60 and over) to children (aged under 15) was more than 185:100, significantly higher than the national average (144:100) (Statistics

**Table 7.1 Migration statistics for Alytus County, 2009 and 2010**

| Total migration to/from Alytus County | | | |
|---|---|---|---|
| | In-migration | Out-migration | Balance |
| 2009 | 2770 | 3843 | -1073 |
| 2010 | 2792 | 8141 | -5349 |
| International migration to/from Alytus County | | | |
| | In-migration | Out-migration | Balance |
| 2009 | 340 | 854 | -514 |
| 2010 | 242 | 5153 | -4911 |

*Source*: Statistics Lithuania

Lithuania 2011a). Accordingly, there were 312 old-age pension beneficiaries per 1,000 working population in Alytus county in 2011, compared to 283 per 1,000 working population nationally. Additionally, Alytus county has higher-than-national-average rates of claimants for incapacity benefits, whilst payments of social assistance benefits to households on low incomes more than doubled in Alytus county between 2009 and 2010 (ibid.).

*Economic Trends and Conditions*

Between 2001 and 2007 the combination of economic growth and out-migration saw unemployment in Alytus county fall sharply to just 3.3 per cent. However, since 2008 the impact of the global economic recession has been felt in Alytus county with a decrease in employment and a significant rise in unemployment. In 2010, there were 67,700 people employed in Alytus county, down by 7,700 since 2008 (or 10.2 per cent). Meanwhile, the unemployment rate increased from 4.1 per cent in 2008 to 15.6 per cent in 2009 and 17.2 per cent in 2010, marginally below the national average (Statistics Lithuania 2011a).

The majority of employees in Alytus county work in the service sector, although agriculture and forestry remain more significant than in many parts of rural Europe, employing over 8,000 people, or around 12.3 per cent of the workforce (see Table 7.2). Occupation in forestry is related to the rich forest resources of the region, whilst agricultural employment is disproportionately concentrated in one municipality. Employment in agriculture, industry and construction all fell between 2009 and 2010, but there was an increase in absolute and relative employment in services (Statistics Lithuania 2011a).

Decreased employment during the recession was followed by a fall in average gross monthly earnings from 1874 LTL (€540) in 2008 to 1692 LTL (€490) in 2010 (Statistics Lithuania 2010). The average disposable household income in Alytus county in 2009 was 803.7 LTL (€233) per month, 15.1 per cent less than in 2008, and significantly below the Lithuanian average of 961 LTL (€278) per month, which itself is low by EU standards. Data from the Survey on Income and Living Conditions shows that average equalized disposable income in Lithuania in 2009 was 3.5 times lower than Luxembourg, and 2.3 times lower than the

**Table 7.2 Employment by sector in Alytus County, 2010**

| Sector | Thousands | Per cent | Change since 2009 (thousands) |
|---|---|---|---|
| Services | 43.0 | 63.5 | +3.1 |
| Industry | 11.5 | 17.0 | -3.7 |
| Agriculture, Forestry & Fishing | 8.3 | 12.3 | -9.1 |
| Construction | 4.9 | 7.2 | -2.9 |

*Source*: Statistics Lithuania

United Kingdom and Ireland, a comparison that has contributed to out-migration to Western Europe for work (Statistics Lithuania 2011b). Gross domestic product per capita in Alytus county is similarly only two-thirds the level of national GDP per capita in Lithuania, which itself is only 58 per cent of the EU average. At the start of the recession, GDP per capita in Alytus county fell by 19.3 per cent in one year between 2008 and 2009, but has marginally recovered since (Statistics Lithuania 2011a). Nearly 60 per cent of regional gross value added (GVA) is generated by the service sector, with agriculture and forestry contributing only 5.1 per cent – though still above the national average. The majority of businesses in the county are based in the urban centre of Alytus city, and over three-quarters have fewer than ten employees. Only 14 businesses in Alytus county employ more than 250 people. The total number of businesses operating in the region has fluctuated during the recession, but increased to 3,362 in 2011.

The recession was also reflected in a sharp decrease in business turnover in Alytus county between 2008 and 2009, which was especially pronounced in the construction, transportation, manufacturing, wholesale and retail sectors, all of which recorded falls in turnover of more than 30 per cent. However, according to the Structural Business Survey in 2010, an increase in business turnover in the county of 9 per cent between 2009 and 2010 suggested signs of a slow recovery (Lithuanian Structural Business Survey 2010). Nevertheless, public confidence has remained low. A survey in November 2009 found that 86.7 per cent of residents viewed their employment prospects negatively or very negatively, and 81.6 per cent assessed the possibility of starting a business as negatively or very negatively (RAIT Market Analysis and Research Group 2010). Reasons cited by respondents for poor economic confidence included a shortage of employers, low wages and poor working conditions (see Figure 7.1), whilst lack of private capital for investment was identified as a major obstacle to business start-ups. The same survey also found that 29 per cent of respondents had at least one unemployed person in their family, with opportunities for reskilling and improving qualifications not thought to be good (ibid.). Although the training infrastructure in Alytus county is well developed, there is recognition that available programmes have not been adapted enough to changing market conditions (Alytus Regional Development Council 2010).

### *Regional Development Programmes*

The main strategic document setting development trends for Alytus county is the Alytus Region Development Plan 2010–20. The plan identifies three priorities – a competitive economy, a cohesive community and a safe, clean and convenient environment in which to live. The objects for the 'competitive economy' priority include improving the business and investment environment, developing tourism and recreation and the cohesive development of the region's countryside. To support a 'cohesive community', the plan aims to improve the quality and accessibility of education, training, health and social services, to increase capacities to adapt to the changing labour market and to develop the infrastructure for culture and sport. Finally, objectives for the third priority area include improving the communications and energy infrastructures, creating and modernizing an eco-engineering infrastructure and developing the region's territory in a sustainable way (Alytus Regional Development Council 2010).

Alongside the Regional Development Plan, rural areas of Alytus county also benefit from the Rural Development Strategy 2010–15. According to this, four rural development strategies have been prepared and implemented by four local action groups using funding from the EU LEADER programme. Whereas the Regional Development Plan is coordinated at the regional scale, the rural development strategies operate separately in each of the rural municipalities, according to their specific needs (Lithuanian Ministry of Agriculture 2012). Additionally, two municipalities – Druskininkai and Lazdijai – belong to the group of 'problematic territories' designated by the Lithuanian Government. These territories have an unemployment rate that is at least 160 per cent of the national average, and a proportion of the population receiving social welfare support that is at least 160 per cent of the national average (Lithuanian Government 2006). These municipalities prepare and implement their own 'Development Programmes for Problematic Territories', the main goals of which are to reduce unemployment and welfare receipts closer to the national average and to raise the quality of life. The programmes include actions aimed at improving the living environment, renovating public infrastructure, stimulating employment, increasing skills and qualifications and optimizing municipal administration (Lithuanian Ministry of the Interior 2011).

## Engagement of Alytus County Businesses in International Networks

As discussed above, the economy of Alytus county, while traditional in character, is also vulnerable to global economic trends. In this section the extent of engagement by businesses in the region in the global economy, and some of the initiatives that have attempted to support the development of international linkages, are discussed.

In common with many parts of Central and Eastern Europe, Lithuania has attracted significant foreign direct investment (FDI) since the liberalization of its economy, and particularly since it joined the European Union in 2004. In December 2010, FDI in Lithuania amounted to over 35500 LTL million (€1000 million), an increase of 12 per cent on December 2008, in spite of the recession. Yet, little of this investment has found its way to Alytus county. FDI in Alytus county accounts for less than 1 per cent of total FDI in Lithuania, a decrease of 13 per cent between 2008 and 2010, against the national trend. In December 2010, FDI per capita in Alytus county was 1940 LTL (€540) (Statistics Lithuania 2011a). Most investment is concentrated in Alytus city, with rural municipalities such as Lazdijai and Varena receiving FDI of less than 300 LTL per capita.

Notable foreign firms investing in Alytus county include Coca Cola Bottling (USA), Stora Enso Timber (Finland and Sweden), Graanul Invest (Estonia),

Litalka Elektronik (Germany) and Sofa Brands (UK). Many of these operations are concerned with manufacturing goods for export within the European Union, such that the region's share of national exports is higher than its share of FDI, though still low. In 2010, exports from Alytus county totalled 839.9 LTL million (€235 million) in value, below the 2008 level but an increase from the trough of the recession in 2009 (Statistics Lithuania 2011). Exports were sent to 60 countries, but three-quarters were to other parts of the European Union with the largest markets in 2009 including Germany (16.4 per cent of all export), Poland (13.9 per cent), UK (9.2 per cent), Russia (9.1 per cent), Latvia (8.1 per cent), Denmark (5.8 per cent.), Sweden (5.3 per cent), Norway (4.4 per cent), France (3.7 per cent) and Italy (3.5 per cent) (Alytus Regional Development Council 2010).

*Business Networks of Rural Firms in Alytus County*

In spite of the above evidence of export activity, the majority of small endogenous businesses in Alytus county have limited engagement with international networks. As part of the DERREG research, an electronic survey was conducted with 55 businesses in Alytus county in 2009, with 15 firms also participating in follow-up interviews. Of the 55 firms surveyed, 50 were independently owned (3 belonged to regional conglomerates and 2 to national conglomerates), and 47 were micro-businesses employing fewer than 5 people. As such, the survey sample was representative of the bulk of Alytus businesses, though it should be noted that attempts to include larger firms more likely to have international linkages met with non-responses, in part due to concerns about commercial sensitivity at a time of uncertain market conditions.

The survey found that 19 of the respondents (35 per cent) claimed not to participate in international markets at all, whilst only four of the surveyed companies claimed to have a 'high' or 'medium high' level of integration with international markets. Considering all transactions and employing the methodology developed for DERREG (see Copus et al. 2011), 78 per cent of the survey respondents were classified as being mostly regional in their orientation, with just 4 per cent classified as fully internationalized (with both international sales and purchases). This was the lowest level of internationalization recorded in any of the DERREG case study regions surveyed (see Table 7.3) (see also Copus et al. 2011). Even if, as other data suggests, the level of full internationalization might be as high as 10 per cent if medium-sized enterprises are included, this is still low by comparative standards.

There are several reasons for the low degree of internationalization of Alytus County businesses. There are only a few sectors in the region where international networks play an important role, including wood products and furniture-making, manufacturing plastic and foam articles, processing frozen berries and mushrooms, the garment industry and dress-making and waste products for animal food. Outside these sectors, the capacity to develop international networks is limited, especially for micro-enterprises. The research suggests that firms need to have

**Table 7.3 Classification of survey firms by degree of internationalization of business activities in Alytus County, compared with other DERREG case study regions**

| Alytus county (LT) | | Övre Norrland (SE) | South Moravia (CZ) | Westerkwartier (NL) | Goriška (SI) |
|---|---|---|---|---|---|
| Sample size | (55) | (47) | (40) | (38) | (20) |
| Mostly regional | 78% | 13% | 37% | 65% | 5% |
| National | 4% | 34% | 5% | 15% | 15% |
| Partly internationalized | 15% | 32% | 33% | 8% | 45% |
| Fully internationalized | 4% | 21% | 26% | 13% | 35% |

certain levels of human resources and accumulated capital to develop successful international cooperation. For the smallest firms, a lack of means for investments, language limitations, small volumes of production, type of production and limited familiarity with overseas markets are key factors restricting their capacity for internationalization. Geographic location is also important. Whilst Alytus county is perceived to be well connected to the larger cities in Lithuania, its international borders are considered to offer limited opportunities. The broader DERREG analysis suggests that proximity to countries with higher purchasing power and higher production costs is a precondition for international trade and cooperation. However, Alytus county's neighbours in Belarus and Poland do not present such possibilities, and except for some specific products (such as high-tech), the relative inaccessibility of the region means higher costs (especially for agricultural and forest products).

The degree of internationalization also varies in terms of different aspects of business activities. Most of the firms surveyed imported some supplies or exported some products, but international collaborations with other firms, accessing international investment capital and contacting international institutions were all unusual. The primary networks that Alytus county businesses have are with other SMEs in the region. These tend to be focused on cooperation around compliance issues and marketing – for example through business and trade associations – and are often based on personal contacts and friendship. Contacts with national organizations and institutions were infrequent, such that Alytus businesses appear to be self-reliant. However, those businesses that do have international relations tend to rate them as more important than domestic contacts.

This pattern of internationalization has shaped the impact of the global economic recession on businesses in Alytus county. A few businesses that had established international networks found that these were weakened by business failures elsewhere. One firm reported that 'all connections with Irish and British enterprises broke down' during 2009, whilst another observed that 'at the beginning ... the company had about 100 foreign partners (in Bulgaria, Slovakia, Czech Republic, Germany, Poland etc.), and now one national partner is left'.[1] However, for most businesses the crisis was felt in terms of its regional impact, especially in terms of reduced demand and customer purchasing power. Yet, only two of the firms interviewed had reduced their workforce due to the recession, and half claimed that the crisis had no impact on their business. Indeed, the survey indicated a slight increase in contacts with customers at the global level, possibly implying a search for alternative markets.

*Network Brokers and Networking Initiatives in Alytus County*

Although the current internationalization of businesses is low, the development of international networks is part of the economic development strategy for Alytus county, and there are a number of institutions involved in offering support of different kinds. In common with other regions studied in DERREG, this support focuses on 'network-brokering' activities, and especially on 'forum facilitation' bringing together groups of firms with a common interest to strengthen trust and foster collective learning and collaboration, perhaps also with regard to international markets (Copus et al. 2011; see also Chapter 2 this volume).

Interviews with institutions involved in network-brokering in the region however revealed that their activities tend to be focused on regional small and medium-sized enterprises, not on rural micro-enterprises, which are perceived to be served by rural development policy measures and local action groups.[2] This distinction reinforces the divide in internationalization observed in the survey. Each of the organizations plays a different role in the regional business support network, but there are some overlaps in activities and functions. The Department of Investments and Foreign Relations of the Alytus District Municipality, for example, promotes networking through its Small and Medium Business Support Fund, partly compensating the expenses of firms participating in fairs, business missions, seminars and courses, as well as for website development (it also advises businesses on available land, realty and land tax compensation). Similarly, the Alytus Branch of the Vilnius Chamber of Commerce, Industry and Crafts organizes business missions and contact fairs, and posts business proposals on its website allowing prospective partners to contact each other. It also provides a safety-net for firms in providing certification to meet requirements for export

1 Quotes from interviews with business owners and managers.

2 Interviews were conducted in late 2010 and early 2011 with the Alytus District Municipality Department of Investments and Foreign Relations, the Alytus Business Advisory Centre, the Alytus Branch of the Vilnius Chamber of Commerce, Industry and Crafts, the Alytus Business Innovation Centre and the Lithuanian Innovation Center Section for Southern Lithuania. Information was also provided by the Alytus Region Business Association.

licences and supplies-targeted information via a database with firms grouped by activity.

The Alytus Business Advisory Centre (ABAC) works as a catalyst for business network development, with its director explaining that, 'when creating any network it is necessary to look for mutual benefit for firms and possibilities for the broker to assist these needs, to help the network develop. It includes market, partner search, possibilities for new products, improvement of quality and innovations'. Initially, firms were engaged through conferences and seminars, but the involvement of businesses now happens through word-of-mouth and firm-to-firm recommendation. The ABAC now works with organizations at local, regional, national and international levels.

The Lithuanian Innovation Center Section for Southern Lithuania and the Alytus Business Innovation Centre are primarily concerned with research and development, training and education. The Lithuanian Innovation Center connects businesses, science and other public institutions, including helping with searches for international partners. In particular, it implements three EU initiatives (EBIN: European Business and Innovation Network; InPuls and InPuls+), aimed at fostering international cooperation in technological innovation and development. The Alytus Business Innovation Centre is more locally focused as a mediator between science and business through Alytus College and regional firms. They organize seminars and prepare projects.

Interviews with managers of these initiatives however identified a number of key challenges faced by network-brokering. Firstly, possibilities for international cooperation are restricted by finance. Electronic networks can provide information, and business networks are usually funded by projects for a fixed period, but there is no regular funding for ongoing physical networking. Secondly, maintaining a network once established involves time and effort. As the director of ABAC noted 'to form a network is not difficult, it is difficult to maintain and develop it further, often due to finished funding. In order to maintain and develop the network by itself at least two years are needed after the end of the project'. Thirdly, potential for international cooperation is sometimes restricted by the apathy of managers and by their insufficient knowledge of foreign languages. According to the LICSL representative 'those who are searching and are brave, persistent and have goals usually achieve them'.

Nevertheless, there are a number of examples of successful international network brokering in Alytus county. These include a business mission supporting integration of Alytus firms into international wood clusters networks; a project promoting cross-border cooperation in the wood enterprises cluster; the international electronic network project for innovative wood and metal clusters, the Lithuanian and Polish International Trading and Business Centre and the Alytus and Balstogė-Suvalkai Regions Business Cooperation Development project aimed at increasing competitiveness in EU markets (Kairytė 2010). What these initiatives have in common is the use of clusters, competency development and the exchange of experience using cross-border cooperation opportunities, as well as the establishment of electronic networks and the use of websites as network operation tools.

In turning to the next interconnected research theme, that is, migrant workers, the remainder of the chapter focuses on the backgrounds and migratory pathways of these individuals, and their social and economic integration into the region.

## Migrant Workers and Return Migrants in Alytus County

Lithuania has the largest net negative migration per 1,000 population in the European Union. In the decade up to 2010, 204,700 people emigrated from Lithuania and only 64,200 migrated to the country (Statistics Lithuania 2011c). The imbalance in Alytus county is even more pronounced. In 2010, over 5,000 people emigrated from Alytus county and fewer than 250 arrived in the region from abroad, making a net emigration rate of 28.8 per 1,000 inhabitants – one of the highest in Lithuania (see Table 7.4). Emigration is driven by poor economic prospects in the county and the prospect of higher earnings abroad, especially in Western Europe. However, the dynamics of migration are not quite as simple as these figures initially suggest. Nearly 80 per cent of immigration recorded to Lithuania in 2010 was comprised of Lithuanian nationals returning from living and working abroad, including some return migrants to Alytus county (Statistics Lithuania 2011b). Furthermore, there is also a small flow of immigrants into Lithuania, especially from countries of the former Soviet Union, some of whom fill labour vacancies created by out-migration.

This section discusses two groups that form part of this migration flow and which were studied as part of the DERREG project: immigrants from the former Soviet Union, and return migrants from Western Europe. Both of these groups connect Alytus county with other parts of the world, and both present challenges for regional integration.

**Table 7.4 International migration to and from Alytus County, 2010**

| | Emigrated | Immigrated | Net migration | Per 1,000 population | | |
|---|---|---|---|---|---|---|
| | | | | Emigrated | Immigrated | Net migration |
| Lithuania total | 83157 | 5213 | -77944 | 25,3 | 1,6 | -23,7 |
| Lithuania total rural areas | 20550 | 952 | -19598 | 18,9 | 0,9 | -18,0 |
| Alytus County | 5153 | 242 | -4911 | 30,3 | 1,4 | -28,8 |
| Alytus County rural areas | 1287 | 67 | -1220 | 18,4 | 1,0 | -17,4 |

*Source:* Statistics Lithuania 2011c

*Insignificant and Shrunken Labour Immigration*

Immigration to Alytus county is very limited and heavily overshadowed by emigration. Only 242 migrants arrived in the region from outside Lithuania in 2010, and of these only 67 moved to rural municipalities – an immigration rate of only 1 per 1,000 inhabitants, compared with an emigration rate of 18.4 per 1,000 inhabitants (Statistics Lithuania 2011c). However, these immigrants are only the latest of a long line of international migrants to the region. In an interview with the leader of Alytus Russian Culture Association 'Malachit', it was suggested that people of the following nationalities live in Alytus Region: Russians, Armenians, Jews, Tartars, Germans, Azerbaijani, Polish, Ukrainians, Byelorussians, Udmurts, Komi, Basques and Georgians. The majority of immigrants are Russians, Russian speakers and Tartars, the others are solitary instances. The Tartars came to the region in earlier historical times whilst Russians came mainly during the Soviet period as part of Soviet labour migration policies.

As part of the research, face-to-face qualitative interviews were conducted with 11 immigrants to the region, all of whom were Russian-speakers, six from Russia, three from Ukraine, one from Belarus and one from Estonia. Most of the interviewees had moved to the region between 1946 and 1987, during the Soviet period; but two had arrived since 2001. Most of the interviewees had either been posted to Alytus county in the Soviet labour market system, or had moved with partners who had been posted to the region. In some cases, the interviewees had moved to join Lithuanian partners.

All the interviewees were retired and dependency on a very low-income pension contributed to their economic marginalization. Seven of the 11 interviewees had monthly incomes of between €200 and €400, and four had monthly incomes of less than €200. The jobs that they had held in the region were mostly low paid with poor opportunities for advancement, but they had provided a degree of stability that the interviewees now missed. As such, financial constraints were identified as the key challenge facing them and their living in the region:

> Too small pension for medicine and accommodation payments (heating, etc.). Otherwise it's nice. [Interview, immigrant 1]

> I don't have sufficient means for the living. The pension is not sufficient for the medicine and other things. [Interview, immigrant 8]

At the same time, almost all of the interviewees mentioned the natural environment and rural landscape as the biggest advantage of living in the region. The second positive factor was leisure and cultural activities; the third, tolerant and hospitable people; and the fourth, accommodation possibilities. Other identified advantages included the cleanliness of Alytus town, good shops and proximity of family and relatives living in the region. The majority of the interviewees felt that they been accepted and tolerated by Lithuanians:

> Lithuanians accepted Russian very well; I did not feel any difference. [Interview, immigrant 6]

> I feel good among the locals, made friends. [Interview, immigrant 6]

The children of the respondents had generally grown up in Alytus county and most spoke Lithuanian and had experienced no problems with integration, even though when the Russian-speaking community was larger it had its own Russian-language secondary school in Alytus city. Help in finding accommodation and jobs on first arriving in the region had generally been provided by relatives and friends, though some had received accommodation from the public administration and one interviewee had received help finding accommodation from their employer. Informal social networks remain important, with most of the interviewees belonging to two organizations, the Alytus Cultural Centre of National Minorities and the 'Bočiai' club for elderly people.

*Patriotic and Forced Return Migrants and Repeat Emigrants*

Whilst immigration of foreign nationals into Alytus county has declined, an important new trend is the return migration of Lithuanians who have been migrant workers in other parts of Europe or elsewhere. Nine such return migrants were interviewed as part of the research in 2010. These migrants left Lithuania to earn money, to gain experience and to obtain a perceived better quality of life, but had returned for a variety of reasons that can be grouped into three main categories.

Firstly, the most commonly cited motivations for returning were health or family reasons, or what might be regarded as 'forced return migration' that is not entirely voluntary but dictated by circumstance. For instance, one migrant who had lived in Armenia spoke of difficult conditions resulting from an earthquake, war and food shortages, but ultimately decided to return to Lithuania having been diagnosed with skin cancer and advised to change climate. Another returned due to their mother's deteriorating health. Secondly, some migrants opted to return for patriotic reasons, feeling that they belonged in Lithuania. Interestingly, this included at least one second generation emigrant:

> I was born in a Lithuanian family (in USA), to which it was very important to save Lithuanian culture – language, traditions, etc. since Lithuania was under occupation. Parents were communicating mainly with Lithuanian society, were dreaming of Lithuania as homeland. So I was practically 'programmed' to return and to be Lithuanian. [Interview, return migrant 1]

> Being Lithuanian, I never felt myself a citizen of any other country except Lithuania, and to my opinion, citizens have to live in their country. [Interview, return migrant 6]

Thirdly, some migrants returned because they had achieved their goals or, alternatively, because anticipated opportunities had not been forthcoming. One interviewee commented that they returned because:

> I wanted to continue my studies in Lithuanian. After working some time in the pub I got tired of the routine, long working hours. I didn't see myself doing this job permanently. (interview, return migrant 5)

However, returning migrants did not necessarily find that their experience increased their ability to find a job in Alytus county. In many cases they had been working in jobs abroad that did not match their professional qualifications or training, such that the opportunity for professional growth had been limited. Some did undertake training abroad to develop new skills or experiences, including in jobs as a computer operator, bar tender, house salesperson, loan provider and a truck driver. Yet, in contrast to experiences observed in some other regions (see Chapter 4, this volume), those trained abroad, and those who had gained new skills abroad in their own professions, often found it difficult to transfer these competences back home due to different 'rules of the game', as one returnee from the United States observed:

> The knowledge that emigrants accumulate abroad (USA) about business and similar is not always applicable in Lithuania – return migrants don't know many peculiarities and circumstances, related to the situation in Lithuania. In general if you start a new business you need to read about everything yourself. The level of private business consultation centres is not sufficient; they are suitable only for people without basic business knowledge. There should be some solution found for this. Other thing there is no clarity about reality. People ask each other if it got cheaper or more expensive, how it will be in future. Not clear what to do and when to do better (in terms of investments). There should be more information provided about that. There is no centralized institution to which return migrant could turn to get answers to all his questions. Those institutions that exist do not provide sufficient information. [Return migrant 6]

Thus, only five of the nine interviewees were satisfied with their actual living situation in Alytus county. They appreciated proximity to family and friends, natural environment, Lithuanian language, good accommodation possibilities and a cheaper cost of living; but they also identified negative features of life in Alytus county such as low wages, unemployment, bureaucracy, poor performance and inefficiency of public officials, competition, rude sales staff in shops, the selfishness of politicians, lower quality of life and the peripherality of the region. The first experiences they had on returning were often related to these negative factors: bureaucracy, inappropriate behaviour of public officials, coldness and bad moods. Alytus county began to look small and peripheral after returning from abroad. Those who were dissatisfied with life in Alytus were considering emigrating again. Overall, five of the interviewees planned to stay in the region, three planned to go abroad again and one was not sure. Two of the interviewees planning to stay said that they would consider leaving, but were restricted by age and family circumstances.

According to the head of the International Organization for Migration Vilnius Office, Audra Sipavičienė (2009), successful return migration is possible only if the political, economic and social situation in the home country is stable and attractive. It also depends, it can be argued, on how much the state is interested in facilitating the return of its citizens. This research suggests that Alytus county has not yet met these conditions for successful return migration, and that even planned returns might be delayed for some time due to the economic recession.

### *Regional Integration through Cultural and Educational Activities*

Although immigration levels in Alytus county are low, there are in the region of ten grassroots organizations and initiatives that seek to cater for immigrants or cultural minorities, or which aim to promote integration. However, several of these are cultural organizations that are no longer primarily constituted by immigrants, but who draw their members from Lithuanians with different ethnic roots, or from those interested in other cultures. Examples include the Russian Saturday School, Alytus Russian Culture Association 'Malachit', Belorussian Cultural Fellowship 'Spadčyna', Varėna District Belorussian Community 'Nadzeja', Druskininkai Subdivision of the Lithuanian Polish Union, Alytus County Tartar Community, Varėna District Lithuanian Tartar Community and the Druskininkai Jewish Community. The majority of more recent immigrants identify with the Alytus Cultural Centre of Ethnic Minorities, whose activities are focused on social and cultural integration.

Collectively, the above organizations promote activities that seek to maintain the cultural life of minority groups and build bridges between Lithuanians and other cultures. These include Russian literature and language classes, concerts, performances, readings and exhibitions and celebrations of Russian festivals and memorable dates. However, only one initiative is primarily focused on economic integration: the 'Entrepreneurship Development and Integration into Business Market of Foreigners who Received Asylum in the Republic of Lithuania' project. This initiative, operated by the public institution Alytus Business Incubator, was launched as part of the programme of the European Refugees Fund in Lithuania in 2008, and included activities related to the creation of businesses and the improvement of employment opportunities for asylum migrants (Kairytė 2010).

The main problem faced by all initiatives, however, is funding. Most of the organizations have entrance or membership fees, and some receive donations, but only three receive public funding, and two, private sector sponsorship. Nonetheless, activities of this kind have become more diverse with broader geographic coverage in terms of cooperation, with opportunities for international cooperation and exchange being explored across the Baltic Region, Karelia and Saint Petersburg.

## Conclusion

The research presented in this chapter shows that Alytus county occupies an uncertain position in the 'global countryside' (Woods 2007). Opportunities for international engagement have been created by economic liberalization since the restitution of Lithuanian independence in 1990, and by membership of the European Union in 2004, yet impacts are uneven at best. In comparison with the rest of Lithuania, Alytus county – and especially its rural municipalities – has performed poorly in attracting foreign direct investment and in building international business networks. It has also suffered from proportionately greater emigration.

International engagement by businesses in Alytus county has been primarily concentrated on medium-sized enterprises, especially in certain sectors. The majority of rural-based micro-businesses – which comprise the bulk of businesses in the region – have few if any international trading relationships. This reinforces their dependence on regional and national markets with their lower spending power, and thus limits their potential for economic growth. These characteristics also increased their vulnerability to the economic recession in 2008–09, with sharp falls in turnover and employment recorded in Alytus county. However, firms strongly dependent on international partners seemed no less secure than regional firms in times of global crisis. In fact while locally oriented firms shrunk, but still functioned, international firms collapsed if they lost their clients. Consequently it appears that mixed firms or those firms with balanced regional and international networks were the most secure.

In terms of the people of Alytus county, limited economic opportunities have restricted prospects for immigration to the region, and in many ways have spurred on emigration, with large numbers leaving to live and work abroad, especially in Western Europe. Some migrant workers have started to return, and while this is a very positive and welcome position it is worrying that many of those who have returned commonly experience difficulties finding appropriate work. Indeed it is equally troubling to witness many of these returnees being drawn into patterns of repeat emigration.

The weaknesses of Alytus county in the emergent global countryside reflect the structure of its economy, the lack of private capital for investment, limited familiarity with international markets of business owners and managers, the ageing rural population, the lack of human resources and qualified labour due to emigration and geographic location. Unlike many border regions, Alytus county does not benefit significantly from its international frontiers with Belarus and Poland. The adjacent regions are not affluent enough to provide business opportunities, the rigidity of the Belarus border as the eastern frontier of the EU deters trade and migration, and longer transportation distances (crossing more borders) increase costs. These locational factors continue to deter potential foreign investors, as does wariness about the geopolitical ambitions of Russia, outweighing the on-paper attractions of the region's low wage economy for transnational corporations.

Nevertheless, the DERREG research has found that there are examples of good practice in Alytus county's response to globalization. These include initiatives to broker international networking and the promotion of the cultural and economic integration of migrants. Moreover there are other signs of positive development: an increase in business transactions at the global scale and economic recovery since 2009, return migrants wanting to contribute to the regional economy and migrants abroad being able to act as intermediaries for international networking. The natural, touristic and recreational resources of Alytus county are also increasingly recognised as important elements in enhancing the region's international development opportunities in the years ahead.

## References


Governor Administration of Alytus County (2011) *About Alytus County*. Available at: http://www.alytus.aps.lt/ava/changeSite.do?siteId=&inlanguage=en [accessed November 2011].

Alytus Regional Development Council (2010) *Alytus Region Development Plan 2010–2020*. Available at: http://www.pvcentras.lt/public/images/global/PROJEKTAI/Alytaus%20RPP_bendras_2010%2006%2014_FINAL%20%283%29.pdf [accessed November 2011].

Copus A. (2001) From core-periphery to polycentric development: concepts of spatial and aspatial peripherality, *European Planning Studies*, 9, 539–52.

Copus A., Dubois A. and Hedstrom M. (eds) (2011) *Deliverable 1.4 Summary of Research Findings for Work Package 1*, DERREG project. Available at: http://www.derreg.eu/content/resource-centre [accessed February 2013].

Eurostat (2010) *The Original OECD Urban-Rural Typology Applied to NUTS 3 Regions*. Available at: http://epp.eurostat.ec.europa.eu/statistics_explained/images/4/4e/The_original_OECD_urban-rural_typology_applied_to_NUTS_3_regions.PNG [accessed November 2011].

Kairytė, E. (2010) *Good practice of regional innitiatives in Alytus county*. DERREG Good Practice Database. Available from: http://www.derreg.eu/content/good-practice-database [accessed April 2011].

Lithuanian Government (2006) *Government of the Republic of Lithuania Decree No. 96 Concerning the Change of Selection Criteria of Problematic Territories in the Decree No. 428*. Available from http://www3.lrs.lt/pls/inter3/oldsearch.preps2?a=270086&b= [accessed November 2011].

Lithuanian Ministry of Agriculture (2012) *Rural Development Programme 2007–2013*. Available at: http://www.zum.lt/lt/zemes-ukio-ministerija/kaimo-pletra/parama-pagal-2007–2013-m--programos-priemones/kaimo-pletros-2007---2013-metu-programa-ir-strategija/ [accessed November 2011].

Lithuanian Ministry of the Interior (2011) *Problematic Territories*. Available at: http://www.vrm.lt/nrp/index.php?id=169 [accessed November 2011].

RAIT Lithuania Market Analysis and Research Group (2010) *Results of Population Poll*. Annex 2 of Alytus Region Development Plan 2010–2020. Available at: http://www.pvcentras.lt/public/images/global/PROJEKTAI/2%20priedas.pdf [accessed February 2011].

Sipavičienė, A. (2009) *Return Migration: Theoretical Insights and Situation in Lithuania*. Vilnius: International Organization for Migration.

Statistics Lithuania (2010) *Counties of Lithuania 2009*. Vilnius: Lietuvos statistikos departamentas.

Statistics Lithuania (2011a) *Counties of Lithuania 2010*. Vilnius: Lietuvos statistikos departamentas.

Statistics Lithuania (2011b) *International Migration of the Lithuanian Population 2010*. Vilnius: Lietuvos statistikos departamentas.

Statistics Lithuania (2011c) *Demographic Yearbook 2010*. Vilnius: Lietuvos statistikos departamentas.

Statistics Lithuania (2011d) *Structural Business Statistics Survey 2010*. Vilnius: Lietuvos statistikos departamentas.

Vagnorienė, N. (2010) *Preparation of Alytus Region Development Plan, 2010–2020*.Aavailable at: www.alytus.aps.lt [accessed November 2011].

Woods, M. (2007) Engaging the global countryside: globalization, hybridity and the reconstitution of rural place, *Progress in Human Geography*, 31, 485–507.

Chapter 8

# Transformation of Rural Slovenia: The Pomurje Region in Search of New Development Paths

Barbara Lampič, Irena Mrak and Irma Potočnik Slavič

## Introduction

Rural areas in Slovenia are multifunctional localities with heterogeneous and mosaic structures, exposed to intensive restructuring and trapped between traditional processes and structures on one side and contemporary development processes of modernization, restructuring and globalization on the other. These pressures, evident across the country, are particularly prevalent in the Pomurje (or Pomurska) Region in northeastern Slovenia. The Pomurje Region has always been a peripheral rural area with severe development obstacles created by its spatial marginalization, declining population, modest overall infrastructure and poor transportation connections. The region has been recognized as the least developed and most economically exhausted in Slovenia, albeit recent development processes are slowly changing this situation.

This chapter presents an interesting perspective on the Pomurje Region by focusing on one of the impacts of globalization, a short but very intense process of amenity migration. While this phenomenon has limited impact on the vitality of villages or in influencing local rural communities, amenity migration has been strongly connected with the attractiveness of the 'rural milieu' and natural amenities of the region. Indeed for the first time in the history of the Pomurje, its rich environmental capital has been recognized as a valuable development resource that could change the traditional perception of this particularly peripheral rural area.

## The Mosaic Geographies of Rural Slovenia at the European Crossroads

According to OECD criteria, the entire Slovenian territory (12 regions at NUTS 3 level covering 20,273 km²) is defined as predominantly or significantly rural. This is reflected in several indicators: the low rate of urbanization (approx. 50 per cent), low population density (only 102 inhabitants per km²), and the prevalence of a dispersed settlement pattern where 90 per cent of the 6,030 settlements in the country have less than 500 inhabitants (SORS 2014). The rural character of Slovenia can also be thought of from a qualitative perspective. Heterogeneous cultural landscapes have strong rural features, reflecting relatively distinctive rural spatial practices that are strongly linked to housing (intensive suburbanization in the last three decades) and consumption and less to production. Overall the image of rural society and the perceived quality of life in rural areas – what Halfacree (2006) calls the everyday lives of the rural – have positive connotations in Slovenian society. However, contemporary analysis at smaller scales (Klemenčič et al. 2008) has revealed the heterogeneous structure of Slovenian rural areas, with variations between different districts. Several relevant factors that have induced and generated these mosaic geographies of rural areas include:

1. *Attractive geographical diversity*: four major European geographical macro-regions are represented in Slovenia: the Alps, the Mediterranean, the Pannonian and the Dinaric-karst macro-regions. Geographical diversity is reflected in very distinctive regional landscapes and geographical structures which change rapidly over short distances, alongside variations in development and socio-economic patterns. Following Kovačič et al. (2000), three types can be distinguished: a) peripheral, hilly, mountainous and border rural areas exposed to population out-migration and other relevant processes of depletion, which cover more than 40 per cent of national territory with nearly 15 per cent of inhabitants; b) significant rural areas with agricultural landscapes primarily, located between suburban areas and the peripheral areas of out-migration, which make up half of the national territory; and finally c) peri-urban rural areas covering 6 per cent of the national territory but hosting nearly 15 per cent of population.

2. *Location at the European crossroads with simultaneous concentration and peripherization processes*: Slovenia, located at the southern fringe of Central Europe, on the edge of Mediterranean and Pannonian Basins and close to the Western Balkans, has long been a crossing point for travel routes. In the last 20 years, the emergence of new pan-European traffic corridors in the enlarged European Union has heightened this significance and consequently impacts the reshaping of rural areas. Moreover, as Slovenia shares 1,200 km of frontiers with Croatia, Italy, Austria and Hungary, it is important to underline that globalization along with EU integration has also brought more intensive networking with neighbouring regions across conventional national borders (that is, 'short-reach globalization' involving 'trans-border networkers' (Woods 2013)). Border regions have been exposed to strong and concentrated restructuring which is mostly evident in land use changes, (sub-)urbanization processes, land use conflicts and so on. At the same time, areas located away from developmental corridors have been confronted by new forms of peripherization (lack of FDI, depletion of services and so on).

3. *The socialist legacy in spatial and regional planning meets new EU-related challenges*: before 1991 Slovenia was the most northwestern republic of Yugoslavia, with the best economic performance and most outward-oriented economy. From the 1970s, intra-regional inequalities had been regulated via a spatial model of polycentric development, creating a working network of regional centres (with

approx. 15–20,000 inhabitants) with certain facilities, ruralization of industry, short-distance daily commuting and maintenance of population in rural areas. Another important political decision which enormously affected the development of rural areas was a law on local autonomy (1994) resulting in the formation of 211 municipalities. However, although there have been visible improvements in 'quality of life' for rural populations (enhancement of services, different types of infrastructure, people's direct participation in various activities and so on) there are also weaknesses in proper municipal management, lack of finances, disorder in spatial and regional planning and high expectations from the state regarding delegated activities. With integration into the EU in 2004, new documents and mechanisms regarding Rural Development Programmes were adopted.

4. *Growing rural local inequalities over the last two decades*: since Slovenia became independent in 1991, regional inequalities have become more apparent despite various counteracting measures that have mostly derived from EU regional policy frameworks, particularly since Slovenia became an EU member state in 2004. Perpar (2012) identifies four groups of rural municipalities using principal component analysis, ranging from the most problematic municipalities, clustered in the north-east of the country; municipalities with employment mostly in industry; municipalities that have a most sustainable orientation; and the best economically developed municipalities (Figure 8.1).

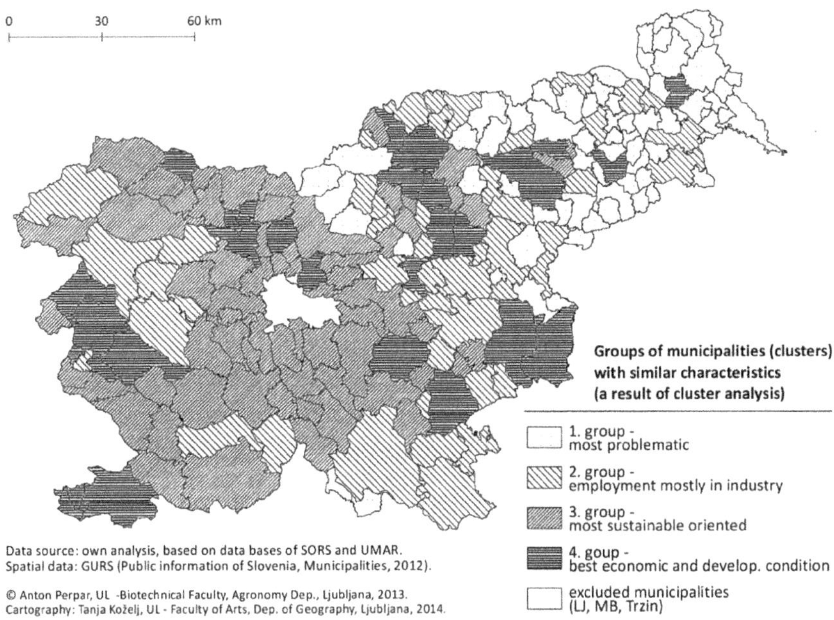


**Figure 8.1 Classification of rural municipalities in Slovenia**

There are several factors that generate regional inequalities, but according to Perpar et al. (2013) economic factors are the most important driver. The economic and development performance of a surveyed municipality is based on the structure of its economic activities: rural areas with a more pronounced active agricultural population and areas with large industrial enterprises are mostly in economic spasm, indicating more developmental problems, long-term unemployment likelihood, modest educational structure, lower entrepreneurial initiative and negative demographic trends. The latter has also been mirrored in the social field and in quality of life. Performance is also affected by the recent administrative organization (fragmentation) of local autonomy with the smallest municipalities territorially being usually the ones with the biggest problems. Municipalities in the first group, with several developmental problems, are consequently characterized by an above average, ageing population, share of active population in agriculture, share of employed persons in public services, unemployment rate (2008–11) and distance from regional centres. Typically these regions have an above average number of registered associations per 1,000 inhabitants and a larger than average share of Natura 2,000 sites. The vast majority of municipalities in the Pomurje Region are facing these structural weaknesses.

## The Geographical Context of the Pomurje Region

The Pomurje Region occupies the northeastern corner of Slovenia, bordering Austria, Hungary and Croatia. With an area of 1,337 $km^2$ (or 6.6 per cent of the country's territory), the region is geographically composed of the flat and fertile Murska Plain (lowlands around the river Mura) with abundant groundwater and agricultural land, and the upland areas of Goričko, Lendavske Gorice and part of Slovenske Gorice. The upland areas in particular are characterized by high biodiversity and a well-preserved cultural landscape (Figure 8.2).

The 117,675 inhabitants represent only six per cent of the population of Slovenia and the population density is below the Slovenian average of 89 inhabitants per $km^2$ (SORS 2014). The region is predominantly rural; with Murska Sobota forming the largest town (population 11,384) and regional centre (SORS 2014). Traditionally the area has been regarded as peripheral – politically, ethnically and economically. Historically, the population depended on farming and seasonal work with agriculture being affected by the legacy of the Hungarian feudal regime, land division and inheritance system.

The Pomurje Region was selected as a study area because of its dynamic historical migration flows as well as its high level of biodiversity and preserved cultural landscape in which settlements are few and dispersed. Processes of recent development in the Pomurje Region can be related to three time periods (Figure 8.3):

**Figure 8.2 Natura 2000 designated areas in Pomurje**

1. Up to 1991: until the end of the First World War, Pomurje formed part of Hungary, in contrast to the majority of modern Slovenia which had been part of the Austrian Empire. As such, the region has always been geographically, politically and ethnically peripheral within Slovenia, with a continuing Hungarian minority population. The regional economy was dependent on agriculture and poor economic conditions encouraged out-migration, with both seasonal and long-term migration becoming regionally recognized as 'normal' economic practice. Before the First World War the migration flow was oriented almost exclusively to the West, with young people emigrating to work and live in North America. In some villages around 20 per cent of the active population left. After the War, Pomurje was integrated into Yugoslavia and the employment agency in Murska Sobota organized seasonal employment in Vojvodina (today an autonomous province of Serbia), France and Germany. At the same time Murska Sobota became a vibrant young urban settlement attracting a work force from the region's hinterland including upland Goričko and Lendavskegorice (Olas 1996b). After the Second World War, population decline was accelerated, the newly reinforced border between Yugoslavia

and Hungary cutting the most important traffic connections and closing increasingly rare employment possibilities. Depopulation was slower in western Goričko where people migrated to work in Austria on a weekly or seasonal basis, legally and illegally (Olas 1996a).

2. From 1991–2004: the reopening of international borders after the decline of the socialist regime in 1991 slowly changed opportunities for trans-border connections. The former Iron Curtain slowly became permeable with an intensification of cross-border traffic, although the sparse population density of areas on each side of the border restricted opportunities for local connections and integration, with most traffic being of long-distance transit in nature.
3. After 2004: with Slovenia entering the EU in 2004 the border with Croatia became the most closed and surveyed border of the region, restricting migration flows to and from the Balkans, whilst the borders with Austria and Hungary were further liberalized. While the border with Croatia was further liberalized following Croatian accession to the EU in 2013, at the time of writing this remains an external border of the Schengen area with strong passport and immigration controls. Pomurje today is an open transit region included in the Pan-European highway network, with these newly formed conditions shaping its future development prospect (Zupančič 2005). One of the first indicators of these new conditions was the rise in amenity migration starting in 2004, with property purchases by foreign homeowners from the United Kingdom, Austria, Germany, Italy and Ireland being attracted to Pomurje as a pleasant and affordable living environment. Foreign property purchases peaked in 2007 and have since declined due to the global economic crisis as well as other globalization impacts.

## Environmental Capital and Generating New Development

The varying permeability of borders and the traditionally peripheral character of the Pomurje Region throughout history have resulted in a relatively well-preserved natural environment and cultural landscape. In general, the environmental capital of the region comprises its biodiversity, natural resources, ecosystem services and (agricultural) land. Each component can have a direct applied value from the aspect of development (economic value), an indirect applied value or no applied value (Lampič and Mrak 2008), depending on the type of development activity and the vision of sustainable regional development. Statistical data for Pomurje reveal that its environment represents the region's major development capital – in terms of its preserved condition and sustainable use. Over 35 per cent of the region's territory is protected within the Goričko Landscape Park (GLP) (IUCN category V) and areas designated under Natura 2000 cover an additional 10 per cent of the Pomurje Region, especially along the river Mura. The GLP

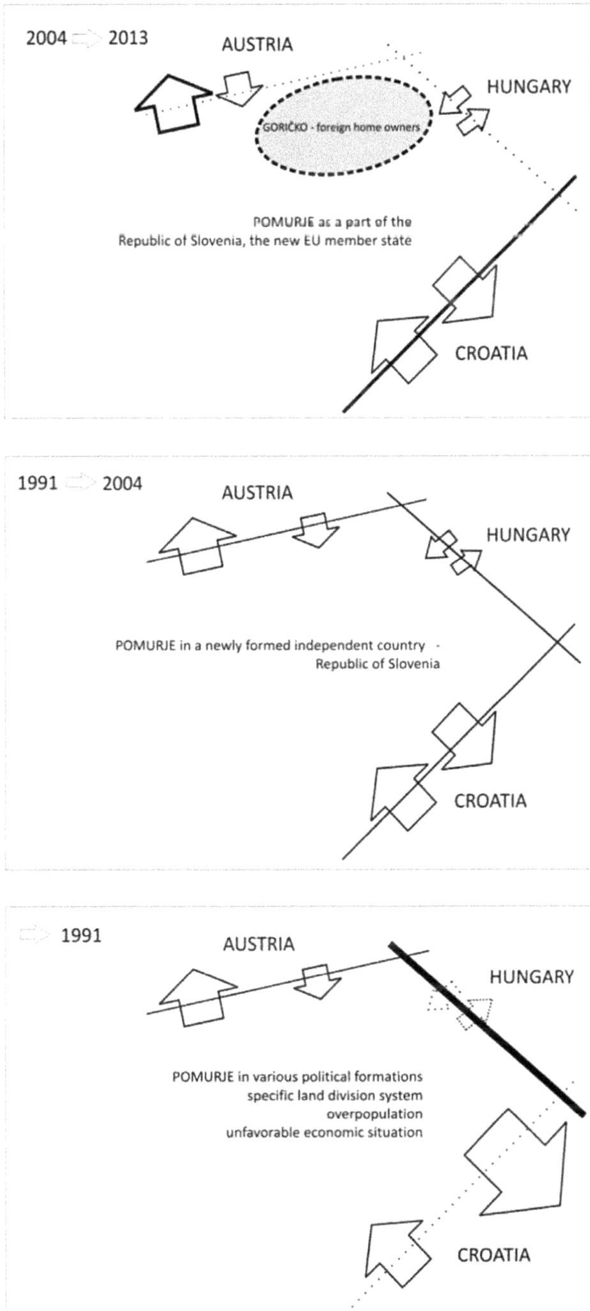


**Figure 8.3 Schematic representation of the geopolitical position of Pomurje**

was established in 2004 as a response to various environmental threats, especially intensive overgrowth due to the abandonment of cultivation. The main objective here was to preserve the typical and traditional cultural landscape of the area and to mitigate socio-demographic decline. The park, the second largest protected area in Slovenia, represents a unique example of historic symbiosis of nature and humans, which still exists to a certain extent at present. Its establishment and development reflect global trends for environmental protection, including transnational cooperation. The GLP is today part of a trilateral cross-border nature park, together with the Raab Nature Park in Austria and the Őrség National Park in Hungary. The importance of preserved environmental capital is also recognized and highly valued nowadays by the main regional stakeholders, regional media and the local population.

Simultaneously, Pomurje is a very important agricultural region in Slovenia. The percentage of agricultural land (61 per cent in the region, 28 per cent in Slovenia) and arable land (42 per cent in the region, 10 per cent in Slovenia) (MAE 2011) is also an important component of environmental capital, positioning the region in the first place in Slovenia in terms of opportunities for food self-sufficiency. With the remediation of environmental problems related to agriculture (water and soil pollution), the region has the potential to become not only self-sufficient but also a region with safe and high-quality food production. The natural conditions of the region have enabled the development of intensive agricultural production on the plains and mostly self-sustained family farming in the upland area of Goričko. Here, traditional farming practices have also contributed to a rich biodiversity as recognized through Natura 2000 areas. As the abandonment of farmland in the region, and especially in the Goričko, reduces biodiversity, the role of protected and preserved areas is crucial. That is, not only do these areas help balance the monitoring and managing of biodiversity and the overall protection of the natural environment, they also support sustainable farming practices and the creation of multifunctional rural areas as well.

Tourism also acts as a catalyst in this process. Tourism in Pomurje has been primarily based on the region's abundant and traditional natural resource – thermal waters – with the most important tourism site in the region being the spa Terme 3000 in Moravske Toplice. It has been developed intensively in the past few decades and is gradually linking its offerings with the region's other natural and cultural amenities. These developments, including the networking of tourism entities, have been supported by the three regional development agencies (the Mura Regional Development Agency, Sinergija, Prleška Development Agency) and the Goričko Landscape Park management institution. Regional Development Agency Mura is the main development initiator in the region, generating and supporting not only project ideas but also offering help in regional stakeholder networking as well as in project management. The role of the other two rural development agencies is localized but they still initiate innovative ideas in the field of both sustainable tourism development and promotion and environmental protection. The agencies are directly involved in sustainable tourism initiatives, as project leaders and more often as project partners, in teams which normally consist of regional, national and quite often international stakeholders. Through their projects, the regional development agencies are indirectly positively influencing the management of protected landscapes, especially in cases where they cooperate with the Goričko Landscape Park. Parallel to the regional development agencies there are 27 municipalities in the Pomurje Region and a comprehensive list of NGOs which are very active in various fields of society. The main objective of all is to achieve equilibrium between the preservation of natural and cultural heritage and sustainable development, favouring agriculture (focused on food production and associated maintenance of the cultural landscape) and tourism based on environmental capital as well as on the rich regional cultural heritage, tradition and strong social capital.

Most of the successful tourism projects (that is, biking trails, the culinary brand 'Scent of Prekmurje', mill trail and so on) are directly linked to regional environmental capital. They were initially supported by EU funds (a small percentage combined with national and local funding) and therefore their existence was in many cases time-limited. There are some examples which grew into regional brands and became self-sufficient after EU financing was completed. The best example is the 'Diši po Prekmurju/Scent of Prekmurje' trademark. The idea here was implemented through a project which was initially supported through the EU programme PHARE in 2005 and has been developed to integrate the region's potential in the fields of quality food and traditional gastronomy. Today the 'Scent of Prekmurje' is a regional trademark connecting the providers of culinary and other gastronomic specialties of Pomurje. It unites local food producers, providers and connoisseurs of quality food with the goal of improving food quality as well as protecting and promoting the culinary specialties of Prekmurje. Its trademark is one of being an open decentralized system allowing new members to join in after they satisfy certain criteria. The trademark can therefore be upgraded, increasing quality as well as widening the list of products on offer. It is promoted through various annual events around Slovenia, which are also used to present and grant quality certificates. As well as these larger events, the Association for the Promotion and Protection of Prekmurje Specialties organizes targeted promotional events (so-called Prekmurje evenings, Days of Prekmurje gastronomy and so on) together with local caterers. There are two products that they already managed to standardize and protect – Prekmurje ham (awarded 'Protected designation of origin' status) and Prekmurje layer pie (recognized as a 'Traditional specialty guaranteed'). The support of the European Agricultural Fund for Rural Areas enabled the implementation of the project 'Promotion of Prekmurje ham and Prekmurje layer pie' in which the two products were presented at gastronomy and tourist fairs, various public events as well as in stores and shopping malls. The trademark unites about 30 certified providers of Prekmurje layer pie (coming mostly from the Pomurje Region; partially from other parts of Slovenia). The group of Prekmurje ham producers and providers is located in the Pomurje Region but is well established in Slovenian, as well as international, markets. The products are well embedded in the overall tourism offer of the region and represent a success story of inclusion of local products into regional tourism offers. They also protect the geographical origin of Prekmurje wines. 'Scent of Prekmurje' also offers other products that are made in Prekmurje and have a unique local character. The 'Scent of Prekmurje' trademark is a good example of an initiative that was originally supported by EU funds and has then developed into a well-recognized and active business which not only contributes to the regional economy but also sustainably influences the regional tradition.

The abovementioned examples are some of the first steps in the regional rediscovery of endogenous potentials and how the establishment of the Goričko Landscape Park represents a unique example of human adaptation to the natural environment. This unique landscape was in decline as increasing numbers of inhabitants gave up their small farms, allowing fields to overgrow and homes and outbuildings fall into disrepair or collapse. The combination of the establishment of the park, and the arrival of amenity migrants after 2004, contributed hugely however to improving this situation.

## 'Unpredictable Response': Amenity Migrants in the Pomurje

Overall development and improvement of transportation infrastructure and connections, active regional competitiveness, globalization and sensitivity to environmental quality gained a leading role in the determination of migration and rural welfare (Chi and Marcouiller 2013). Increasing attractiveness of peripheral rural localities is also connected with lower property prices, the recognized high natural amenity of rural areas and nostalgia (Woods 2007b). These factors were very significant in the growing market for foreign property investment in rural regions of Central and Eastern Europe. Amenity migrants are strongly attracted by pleasant natural environments and favourable economic conditions, and tend to avoid large metropolitan areas (Walters 2002). Most commonly they are found in isolated areas valued for their 'untouched' natural or rural amenities. King and Patterson (1998) in their research on the British community in Tuscany defined various motives for migrating, including an escape from urban environments, attractions to climate and landscape and viewing a region as a place to where they are able to escape from certain perceived negative aspects of the contemporary world.

Benassi and Porciani (2010) define two groups of amenity migrants. Retirement migrants, who may have a history of international mobility or tourism and who tend to move with their spouses and generally have high levels of education and, secondly, those people who want (literally) to 'get off' or 'get out'. These are typically middle- or upper-class people who make the decision to alter/abandon a previous lifestyle. Foreign homeowners in Slovenia mostly belong to these two groups, that is, individuals from economically developed countries (UK, Germany, the Netherlands, Italy, Austria), usually from urban centres, looking for a different lifestyle (Lampič and Mrak 2014).

Amenity migration cannot be understood as a simple movement of people, as it has a variety of impacts which can reshape the character of environment and natural resource amenities. These migrations often initiate a shift from production-oriented activities (farming, forestry and so on) to multifunctional land uses (that is, organic or low impact agricultural practices, protection of native ecosystems, intensive recreational uses (Abrams et al. 2012)). The amenity migration flows in Slovenia were examined using data related to real estate purchases. Property acquisition by foreign owners is a new phenomenon in Slovenia and coincides with Slovenian integration into the EU in 2004. Following this integration, legislation change made it easier for those outside Slovenia to buy property in Slovenia. Invariably some were critical of this move, fearing significant property purchases would be made by foreigners although, it can equally be recognized, foreign home ownership (including by both second homeowners and permanent migrants) can also be considered a positive development. That is, foreign home ownership can have a stimulating effect on the development of the local economy, increase the quality level of services and help renew abandoned villages and cultural landscapes. Indeed this process has been recognized on a smaller scale in many other rural regions of Slovenia. Over the period from May 2004 to the end of 2012, a total of 4,177 foreigners purchased properties all over Slovenia with the highest number of foreign home purchases occurring in 2006 (altogether 740) and 2007 (730) (Tax Administration of the RS 2013). The majority of new real estate owners are second homeowners, but in some regions they have also settled down permanently.

Unsurprisingly, the largest proportion of foreign property purchases has been realized in the Obalno-kraška region on the Adriatic coast (25 per cent). In this region a very important group of foreign second homeowners and cross-border migrants is present. This so-called 'cross-border residential mobility' refers mostly to Italians who live in Slovenian rural areas and commute daily across the border to work in Trieste and its surroundings. There have also been notable numbers of purchases in the Gorenjska (11 per cent) and Goriška (14 per cent) regions. Less expectedly, the region with the second highest volume of foreign property purchases was the Pomurje Region, with 18 per cent of all foreign property purchases in Slovenia in the period from 2004 to 2012. The overall number of foreigners in Pomurje is relatively small (1099 in 2011) (SORS 2013), but the share of the population who are citizens of other EU countries ranks first (28 per cent) among all Slovenian regions (SORS 2013). The purchase of a second home or permanent residence by predominantly British migrants (mostly in the upland area of Goričko) has been attributed to their nostalgic desire for a bucolic rural lifestyle, essentially reflecting the motive for property purchase in remoter parts of rural Europe in terms of the quality of the environment and a simpler way of life. Two other factors have also played a crucial role: the low cost of housing and the presence of low-cost airline connections (for example, London–Graz). Furthermore, in the case of Goričko, a significant catalyst was the role played by a British ex-pat living in the region who started to advertise (via the internet) the idyllic countryside of Goričko to potential UK amenity migrants in 2004 and 2005. This led to the successful purchase of several run-down houses in the region by British owners.

As already mentioned, foreign property purchases in Slovenia have fluctuated, peaking in 2006 with 740 sales, but dropping to 263 sales in 2010 (Tax Administration of the RS 2013). This decrease has been particularly pronounced in the Pomurje Region, with a 77 per cent drop in sales between 2007 and 2010. Since 2010, while Slovenia has remained in an economic crisis, the recorded sale of real estate to foreigners has increased. Such an increase in residential property transactions suggests a revival of the 'foreign property market', albeit its future development and the potential for renewed international sales remain uncertain.

From May 2004 until the end of 2012, a total of 729 foreigners purchased properties in Pomurje. While in raw numbers this total may appear modest, the fact that anyone 'from the outside' was purchasing property in Pomurje made a significant impact on the region and its people. It should be noted that the share of permanent inhabited dwellings per 1,000 inhabitants is lower in Pomurje than elsewhere in Slovenia, with over 11 per cent of the region's dwellings permanently uninhabited. Immediately before Slovenia joined the EU in 2004, over 2,500 uninhabited dwellings were recorded in the region (Pavlin 2009). These were largely homes that were saleable and furthermore their asking prices were quite low. During the period 2004 to 2008, the average price per $m^2$ of private homes in Pomurje was just 56 per cent of the national average (ibid.). By nationality, foreign purchasers of property in Pomurje were led by UK citizens (half of the total), followed by significant numbers of Austrians, Germans and Italians.

The difficulty in projecting future trends regarding foreign amenity migration depends on the influence of a number of complex factors. These can be observed from interviews conducted with a sample of foreign property owners in the region that explored reasons for migration or second home purchases, obstacles encountered in purchasing real estate, experiences of settling down in a new country and of integration into the rural community. The interviews revealed the overall contentment of foreigners with the beautiful and pleasant landscape, new culture and people on the one hand, but on the other, disappointment with the 'formal part' of their inclusion in their new community. This included bureaucratic obstacles and the very poor English language knowledge of public servants. Overall, however, integration into the local community was mostly evaluated as a very positive experience. In the case of our surveyed region the acceptance of foreign homeowners by the locals into their social activities and everyday life was positively appraised.

The high share of in-migrants from EU countries, the poor economic situation, the border and peripheral position of the rural region are all factors which have influenced the different attitude of local people towards new settlers. In the case of the Pomurje Region the increasing trend of international residential property purchases as a result of globalization is recognized as a positive experience particularly in terms of integration with local communities. Foreign homeowners in selected Slovenian regions (including Pomurje) have reached numbers that enable them to have significant socio-economic and cultural impacts on the regions where they have relocated. Given that the costs of living in Slovenia are similar to costs in more economically developed EU member states, combined with the ongoing impact of the current global economic downturn, a new wave of immigration into the country over the next few years is doubtful (Lampič and Mrak 2012).

At present, agriculture in the area remains intense in the fertile plains of Pomurje, whereas the hilly Goričko is a landscape park where traditional use of land is required and encouraged. One of the largest agricultural enterprises in the area, Panvita, owns and rents 3,800 ha of arable land (some in the border area of Hungary) and yet, according to their development strategy and marketing, their need for land is much higher. Nevertheless they are primarily focusing on the plains of the region and not on the hilly part where industrial farming is limited due to the natural conditions. Overall the competition for land in the plains of Pomurje is severe while being less present in the Goričko hills. Therefore, according to the interviews, the pressure on the environmental capital of the region is mostly present in the vicinity of the Mura river, proclaimed designated area of Natura 2000, and less in the Goričko Landscape Park.

## Rural Transformation and the Need for Adaptation, Modernization and Joint Governance

The development of rural areas in Slovenia, in comparison with rural development in other more economically developed countries in Europe, is unique in many ways (Klemenčič et al. 2008). In the post-Second World War period rural Slovenia was shaped by a political system that strongly emphasized industrial development and suppressed development of the agrarian sector. Legal frameworks, along with motorization and the dispersion of industry around the country, caused rapid and mass social restructuring and the emergence of part-time farmers. In the 1990s and 2000s, after the disintegration of Yugoslavia, establishment of the independent nation state and EU integration (2004), a new development era began for Slovenian rural areas. This era has been characterized by the intersection of two basic developmental patterns: a) the shortcomings of rural development up until that time (unsuitable land tenure structure of farms, a large share of agricultural land in the hands of non-farmers, overgrowth of arable land, a high share of elderly people in the population, rural poverty, decline of social ties in the villages, development weaknesses of extensive rural areas) and, b) new development aspirations (suburbanization, significantly improved quality of life in rural areas, retired returnees, new domestic and foreign amenity migration, the advantages of modern technologies, accelerated flows of people, goods and capital and so on). Slovenian rural areas have thus found themselves in a new developmental phase, in which they have been opened up to a wide range of users in a sectoral as well as spatial sense. New developmental structures are taking shape, new initiatives have been fostered and several challenging expectations of rural populations have emerged which require suitable modern conditions (infrastructure, municipal services, qualitative supply provision, attractive cultural landscape and so on) and governance. These repeatedly encountered and newly generated disparities indicate the necessity of advanced operational interfaces to build a partnership approach to rural development: including local initiative, knowledge providers and supportive policies (Wellbrock 2013) which should lead towards a differential approach to rural development shaped by the endogenous potentials of each region and the fundamental opportunities and threats deriving from globalization. Whereas a passive approach (development grants and targeted investments especially in the field of rural entrepreneurship) may be sufficient for more dynamic rural regions, for declining areas a more active approach is required, in which the state and its expert advisory services together with locals participate actively in the development of rural areas.

For several decades, Pomurje was perceived as a typical peripheral rural region. Its remote border position, poor traffic connections and population decrease resulted in various development obstacles. More recently the status and role of the region has been slowly changing: depopulation has slowed down and the educational structure of the population has been steadily improved. The construction of a highway linking the Pomurje to Ljubljana and Graz in 2008 brought the region closer to development centres. Further, there has been increasing recognition of the cultural and economic value of the region's environment and natural resources, and the opportunities these present for endogenous development. Even agriculture has been boosted by a renewed emphasis on food security. The 'lagging behind' situation has been changing but, due to a strong dependence on the industrial sector which itself has been hugely affected during the economic crisis, the process of reconstruction and improvement has proceeded at a slow pace.

Environmental capital has major endogenous development potential in Slovenian rural areas, recognized through various forms of protected areas and areas of Natura 2000. A well-preserved environment along with a rich cultural heritage and strong social capital represent the main pull factors for new rural developments. However, local populations have not recognized all the advantages and benefits protected areas present. Conversely, there is a tendency to see limitations and obstacles for their 'idea' of development. To counteract these perceptions there needs to be cooperation between local populations, different civil initiatives, local, regional and national authorities during the process of designation. Preserved natural environments can be a stimulus for selected globalization processes such as amenity migration and Pomurje is one of the best examples in Slovenia where foreign homeowners support this role in a traditional and environmentally rich rural milieu. While migrants who decided to either settle in the region permanently or use their property as second homes greatly appreciate the preserved rural environment and the openness of the local population on one hand, they also had to deal with unfriendly experiences in dealing with public authorities and institutions in the formal part of their integration. This demonstrates that traditional and peripheral rural areas are not fully ready to embrace on a formal level this new social and spatial phenomenon that has emerged. Amenity migrants in the Pomurje Region have positively contributed to the preservation of cultural heritage by buying run-down houses which they have renovated in the traditional style and in keeping with the region's environment. Consequently it must be recognized that with their sympathies and positive attitude to the 'rural milieu' and local tradition (Cadieux and Hurley 2009), they have also made a positive impact by strengthening the values of the local population, which prior to this had been eroding.

The transformation of the rural landscape is a process driven by individuals who are motivated by economic factors (that is, profit, growth), overall development or preservation and promotion of an area. The need for employment or a lower cost of living (Mitchell and de Waal 2008) also proved to be among the most important factors of amenity migration to Pomurje. Due to the fact that the cost of living in Slovenia is rising, economic factors, in the case of amenity migration, are expected to be less important with potentially the preserved environment becoming the future dominant pull factor.

**Acknowledgement**

We are thankful to all the people of the Pomurje Region who helped us during this research.

**References**

Abrams, J.B., Gosnell, H., Gill, N.J. and Klepeis, P.J. (2012) *Re-creating the Rural, Reconstructing Nature: An International Literature Review of the Environmental Implications of Amenity Migration, Conservation and Society*. Available at: http://www.conservationandsociety.org/text.asp?2012/10/3/270/101837 [accessed: 11 December 2013].

Benassi, F. and Porciani, L. (2010) The dual demographic profile of migrants in Tuscany, in Salzmann, T., Edmonston, B. and Raymer, J. (eds) *Demographic Aspects of Migration*, pp. 209–26. Available at: http://www.springerlink.com [accessed: 3 April 2011].

Cadieux, K.V. and Hurley, P.T. (2009) Amenity migration, exurbia, and emerging rural landscapes: global natural amenity as place and as process, *GeoJournal*, 76 (4), 297–302.

Chi, G. and Marcouiller, D.W. (2013) In-migration to remote rural regions: The relative impacts of natural amenities and land developability, *Landscape and Urban Planning*, 117, 22–31.

DERREG (2011) *European Policy Briefing. Developing Europe's Rural Regions in the Era of Globalization (DERREG)*. Available at: http://www.derreg.eu/content/resource-centre [accessed: 16 April 2011].

Halfacree, K. (2006) Rural space: constructing a three-fold architecture, in P. Cloke, T. Marsden and P.H. Mooney (eds) *Handbook of Rural Studies*, pp. 44–62. London: Sage.

King, R. and Patterson, G. (1998) Diverse paths: the elderly British in Tuscany, *International Journal of Population Geography*, 4, 157–82.

Klemenčič, M.M., Lampič, B. and Potočnik Slavič, I. (2008) Življenjska (ne)moč obrobnih podeželskih območij v Sloveniji, GeograFF, 3, Znanstvena založba Filozofske fakultete, Oddelek za geografijo, Ljubljana.

Kovačič, M., Gosar, L., Fabija, R., Perpar, A. and Gabrovec, M. (2000) *Razvojno-tipološka členitev podeželja v Republiki Sloveniji.* Biotehniška fakulteta, Oddelek za agronomijo, Inštitut za agrarno ekonomiko (Biotechnical Faculty, Agronomy Department, Institute of Agricultural Economics), Ljubljana, 112–4.

Kriszan, M., Burdack, J., Frys, W., Lampič, B., Mahon, M., McDonagh, J., Mrak, I., Nienaber, B., Novotná, K., Trnka, P. and Woods, M. (2011) *Deliverable D 3.5, Environmental capital and sustainable rural development. Report of the EU research project 'Developing Europe's Rural Regions in the Era of Globalisation'.* Available from at: http://www.derreg.eu/content/resource-centre [accessed: 20 September 2011]

Lampič B. and Mrak I. (2008) Vrednote, vrednosti in razvojni potenciali območij varovanja, Dela 29, 145–59.

Lampič B. and Mrak I. (2014) *Amenity Migrants in Slovenia: Between Tradition and Globalization*, 2nd volume on Global amenity Migration (in print).

Lampič, B., Mrak, I. and Potočnik Slavič, I. (2012) The impacts of globalization in rural areas of Slovenia: examples from the Pomurje and Goriška regions, *Geografski vestnik/Geographical Bulletin*, 84(1), 151–62.

Ministry of Agriculture and the Environment (MAE) (2011) *Agriculture*. Available at: http://www.mko.gov.si/si/delovna_podrocja/kmetijstvo/kmetijska_zemljisca/ [accessed: 21 March 2014].

Mitchell, C. and de Waal, S. (2009) Revisiting the model of creative destruction: St. Jacobs, Ontario, a decade later, *Journal of Rural Studies*, 25, 156–67.

Müller, D.K. and Hall, M.C. (2004) The Future of Second Home Tourism. Tourism, mobility, and second homes, *Aspects of Tourism*, 15, 273–8.

Nienaber, B., Frys, W., Burdack, J., Copus, A., Farrell, M., Hedström, M., Kahila, P., Kairyte, E., Kriszan, M., Lampič, B., Mahon, M., McDonagh, J., Mrak, I. and Nadler, R. (2011) *Deliverable D 2.4, International mobility and migration of rural population. Report of the EU research project 'Developing Europe's Rural Regions in the Era of Globalisation'.* Available at: http://www.derreg.eu/content/resource-centre [accessed: 20 September 2011].

Olas, L. (1996a) *Goričko. Regionalnogeografska monografija Slovenije, 9. Severni Subpanonski svet.* Geografski inštitut ZRC SAZU, Ljubljana.

Olas, L. (1996b) *Prekmurska ravnina.* Regionalnogeografska monografija Slovenije, 9. Severni Subpanonski svet. Geografski inštitut ZRC SAZU, Ljubljana.

Pavlin, B. (2009) Trg bivalnih nepremičnin v Pomurju 2004–2008, in Kikec, T. (ed.) *Pomurje, trajnostni regionalni razvoj ob reki Muri, 20. zborovanje slovenskih geografov, Ljutomer – Murska Sobota*. Pomurje: trajnostni regionalni razvoj ob reki Muri: zbornik, 260–272.

Perpar, A. (2012) Pojasnjevanje razvojnih razlik na podeželju, in Potočnik Slavič, I. (ed.) *Potenciali dopolnilnih dejavnosti in podjetništva na podeželju*. Oddelek za geografijo Filozofske fakultete Univerze v Ljubljani, Ljubljana, 81–89.

Perpar, A., Kastelec, D. and Udovč, A. (2013) *Pojasnjevanje razvojnih razlik v Sloveniji s pomočjo multivariatnih statističnih metod, in Nove razvojne perspektive*. Založba ZRC, Ljubljana, 11–22.

Snepenger, D.J., Murphy, L., O'Connell, R. and Gregg, E. (2003) Tourists and residents use of a shopping space, *Annals of Tourism Research*, 30(3), 567–80.

Statistical Office of the Republic of Slovenia (SORS) (2013) *International migration by citizenship, statistical regions, Slovenia*. Available at: http://pxweb.stat.si/pxweb/Dialog/varval.asp?ma=05I2004E&ti=&path=./Database/Demographics/05_population/25_Migration_Change/10_05I20_Migration_Change/&lang=1 [accessed: 17 January 2013].

Statistical Office of the Republic of Slovenia (SORS) (2014) *Demography and Economy Portal*. Available at: http://www.stat.si/ [accessed: 21 March 2014].

Tax Administration of the Republic of Slovenia (2013) *Data of foreign home purchases for the period of 2004–2012*, Ljubljana.

Taylor, P.J. (2004) Regionality in the World City Network, *International Social Science Journal*, 56. DOI: 10.1111/j.0020–8701.2004.00499.

The European Interest (2007) *Succeeding in the age of globalization*. Available at: http://eur-lex.europa.eu/LexUriServ/site/en/com/2007/com2007_0581en01.pdf [accessed: 3 February 2014].

Walters, W.H. (2002) Place Characteristics and Later-Life Migration. *Research on Aging*, 24(2), 243–77. Available at: http://roa.sagepub.com/content/24/2/243 [accessed: 4 April 2014].

Wellbrock, W. (2013) *Well-working Operational Interfaces. A key to more collaborative modes of governance*. Wageningen University, Netherlands.

Williams, A.M., King, R. and Warnes, A.M. (1997) A place in the sun: international retirement migration from northern to southern Europe, *European Urban and Regional Studies*, 4, 115–34.

Woods, M. (2007a) Engaging the global countryside: globalization, hybridity and the reconstruction of rural place, *Progress in Human Geography*, 31(4). London. DOI: 10.1177/0309132507079503.

Woods, M. (2007b) Attractive Ruralities? Re-thinking European peripheries in the global countryside. Rural Society of European Peripheries. *Forum Ifl*, 7, 23–38. Leipzig: Leibniz-Institut für Länderkunde.

Woods, M. (2009) *Deliverable 5.1 Literature Review and Conceptual Framework. Report of the EU research project 'Developing Europe's Rural Regions in the Era of Globalization'*, Aberystwyth. Available at: http://www.derreg.eu/content/resource-centre [accessed: 20 September 2011].

Woods, M. (2011) *Rural*. London: Routledge.

Woods, M. (2013) Regions engaging globalization: A typology of regional responses in rural Europe, *Journal of Rural and Community Development*, 8 (3), 113–26.

Zupančič, J. (2005) *Geopolitične in kulturnozgodovinske razsežnosti ozemlja med Muro in Rabo. Geopolitical and cultural-historic dimensions of the area between Mura and Raba river*. Prekmursko narečje, regionalna zavest in narodna pripadnost. Prekmurska narečna slovstvena ustvarjalnost, Murska Sobota, 20–33.

# Chapter 9

# Towards a Sustainable Regional Economy? The Oberlausitz Region in Transformation

Joachim Burdack, Michael Kriszan and Robert Nadler

## Introduction

The Oberlausitz is a border region located in the south of eastern Germany. It shares borders with the Czech Republic in the south and Poland in the east. This border has become more and more permeable since the political changes of 1989/90 and especially since the accession of Poland and the Czech Republic to the European Union in 2004. Cross-border cooperation helps reduce the negative effects of this peripheral location within Germany in the form of a Euroregion (Euroregion Neisse-Nisa-Nysa) and other cross-border activities of regional actors are well established. During the last decades the Oberlausitz Region has been affected by German reunification and also by three major socio-economic processes: globalization, Europeanization and post-socialist transition. These overlapping processes have significantly affected demographic development, the regional economy and the natural environment, bringing opportunities and challenges for the development of the region, including:

- The traditional textile industry breaking down due to international competition and the employment base in mining and energy production being greatly reduced;
- A massive out-migration of the workforce beginning shortly after German reunification in 1990, and
- Global environmental issues becoming more and more important, offering new opportunities in the field of renewable energy production but also challenging traditional mining activities.

This chapter gives an overview of the socio-economic characteristics of the Oberlausitz Region and presents regional distinctions as well as current trends and developments. Two main challenges for regional development, both a consequence of an increasing global competitiveness, are discussed in detail: first, the attraction and integration of international migrants to maintain a sufficient labour stock for the future development of the regional economy and, second, the realignment of the Oberlausitz Region towards a sustainable use of its natural resources in order to gain independence from the global energy market and be prepared for a post-lignite era. To illustrate how regional actors deal with challenges and try to benefit from opportunities arising from globalization, some current functioning initiatives from the Oberlausitz Region are presented in the final part of the chapter.

## The Oberlausitz Region: Spatial Structures and Developments

The Oberlausitz Region (Upper Lusatia-Lower Silesia Region) forms the eastern part of the federal state of Saxony in Germany and is located in a border triangle with the Czech Republic to the south and Poland to the east. The Oberlausitz consists of the two counties (*Landkreise*), Bautzen und Görlitz, and covers a total area of about 4,500 $km^2$. The region is home to 592,000 inhabitants (as of December 2011, Statistisches Landesamt des Freistaates Sachsen 2012a). Sixty per cent of the population live in peripheral rural areas and a further 35 per cent in rural areas of higher population density. Only 5 per cent of the population live in the suburban part of the Dresden urban region in the western part of the Oberlausitz.

The Oberlausitz Region can be divided for heuristic purposes into five geographic zones or sub-regions, most of which stretch out in an east-west direction (see Figure 9.1):

The Hill and Mountain Area (I) in the south of the region stretches along the border to the Czech Republic and includes the mountain ranges of the *Zittauer Gebirge* and the *Oberlausitzer Bergland.* The sub-region is a traditional centre of the textile industry. The industrialisation of many villages and small towns led to a relatively high population density in this area. The Hill and Mountain Area has considerable potential for tourism, with Oybin and Jonsdorf in the *Zittauer Gebirge* which are especially well developed as tourist centres. The Central Zone (II) is the main transportation corridor of the Oberlausitz along the *Autobahn* A4 and the rail road line between Dresden and Görlitz, and the important historic trade road of the *Via Regia Lusatiae superioris* also passes through this area. The main urban centres of Görlitz and Bautzen are located here, but the Central Zone also includes the *Oberlausitzer Gefilde*, an agricultural area of rich loess soils that extends between Kamenz in the west and Löbau in the east. The Heath and Pond Area (III), a flat zone of low population density, extends north of the Central Zone. A large portion of the area is included in the natural protection area of the UNESCO Biosphere Reserve 'Upper Lausitz Heath and Pond Landscape' (*UNESCO-Biosphärenreservat Oberlausitzer Heide- und Teichlandschaft*) that was instituted in 1996 (cf. UNESCO-Biosphärenreservat Oberlausitzer Heide- und Teichlandschaft 2013). Commercial pond-farming (aquaculture) plays an important role in the sub-region. The Mining Area (IV) in the north of the region constitutes the Saxon part of the larger Lausitz lignite fields that extend into neighbouring Brandenburg. The cities of Hoyerswerda and Weißwasser are the centres of the lignite mining area. Finally the Suburban Area of Dresden (V) includes the most western part of the Oberlausitz. The communities in the vicinity of Dresden were heavily influenced by suburbanisation processes of the Saxon

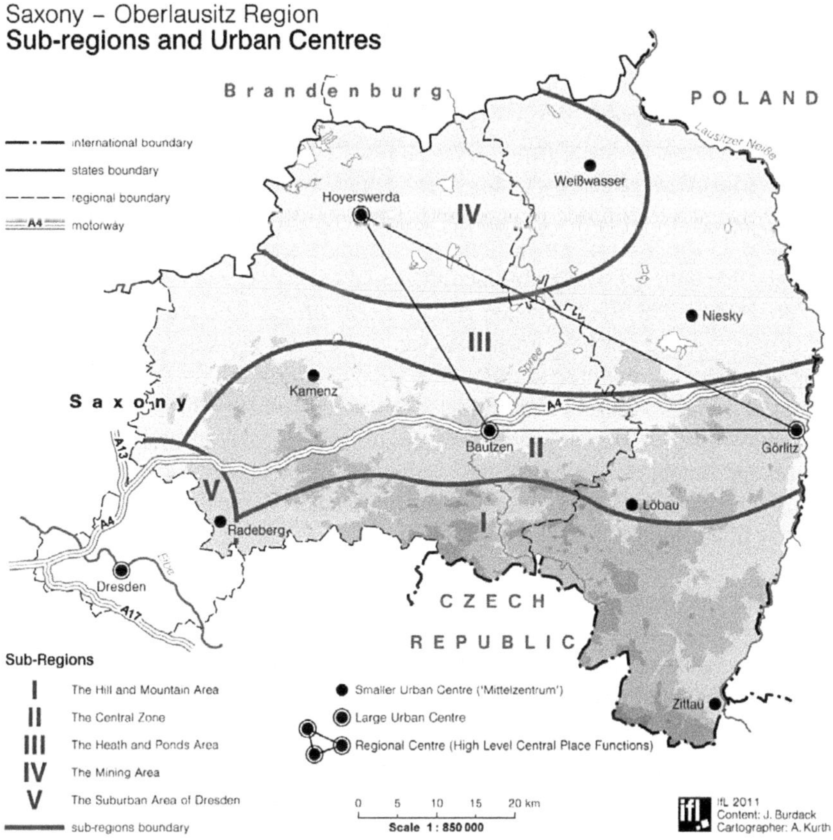


**Figure 9.1 Oberlausitz region: sub-regions and urban centres**

capital. Communities like Ottendorf-Okrilla and Wachau grew by more than 15 per cent in population between 1990 and 2010. The suburban area of Dresden is the only sub-region of the Oberlausitz that did not experience significant population loss after 1990.

*Historic Development of the Region*

The Oberlausitz became part of Saxony in 1635. After the Napoleonic Wars in 1815, the eastern part around Görlitz was attached to Prussia. In 1945, the River Neisse became the eastern border of Germany and the Oberlausitz was now in a peripheral location in a border triangle with Poland and Czechoslovakia (now Czech Republic). The peripherality of the region was accentuated by the fact that the German population was expelled from the Polish part of Silesia, east of the Neisse River. They were replaced by Polish refugees from former eastern Poland.

The ethnic German population that had lived on the Czech side (*Sudetendeutsche*) were also forced to leave their homeland. This meant that the Oberlausitz had in effect new neighbours to the east and south and the ties to the bordering regions were severely disrupted.

The founding of the German Democratic Republic (GDR) in 1949 and the membership of the GDR in the Council for Mutual Economic Assistance (Comecon) in 1950 meant a new phase of economic development in the Oberlausitz Region, especially in the form of intensified lignite mining and processing in the region. Though the centre of the mining activities and the associated energy production was located in the Niederlausitz, that is north of the Oberlausitz Region, the industry has left its mark on Eastern Saxony. The GDR lignite industries required a large number of qualified workers that were drawn to the region from all over the GDR because of good employment opportunities (Friedrich 2003). Towns like Weißwasser or Hoyerswerda grew enormously in size. The mining industry and related branches like machinery, maintenance or transport occupied a large share of the work force and dominated the regional economy along with the textile industry in the south.

### *Settlement and Demography*

There is no dominating primary city in the Oberlausitz, with three major urban centres of comparable importance: Görlitz (55,400 inhabitants), Bautzen (40,500 inhabitants) and Hoyerswerda (36,700 inhabitants) (as of December 2011, Statistisches Landesamt des Freistaates Sachsen 2012b), albeit Hoyerswerda is increasingly losing ground to the other two centres. Regional planning policy designated all three cities as 'partial upper level central places' (*Oberzentren*), that together function as the regional centre (Freistaat Sachsen 2003). Smaller urban centres that function as 'mid-level central places' (*Mittelzentren*) are Kamenz (16,800 inhab.), Löbau (15,900 inhab.), Niesky (9,900 inhab.), Radeberg (18,400 inhab.), Weißwasser (18,600 inhab.) and Zittau (27,800 inhab.) (as of December 2011, ibid.). The Oberlausitz Region has experienced significant population losses in the last decades. The population decreased from 761,700 to 592,000 or -22.3 per cent between 1990 and 2011. Population projections estimate a further decline to 505,600 (-14.6 per cent) by 2025 (Scheibe 2011).

All the major cities of this region suffered population losses in the last 20 years. Hoyerswerda as the centre of the mining area was especially affected and lost almost half of its population (Kröhnert et al. 2005). Population shrinkage affected all sub-regions of the Oberlausitz except for the suburban area of Dresden. This regional population decline is due both to birth deficits and migration losses. Whilst the deficit of births against deaths has been fairly constant over the last 20 years, the migration component shows much more variance. Extreme migration losses in the early 1990s were followed by slight net migration gains in the mid-1990s. Since the year 2000 the migration balance is negative again. The first wave of out-migration after the reunification of Germany in 1990 was basically caused by population groups of working age that were seeking employment opportunities in the urban centres in West Germany (Friedrich and Schultz 2007; Statistisches Landesamt des Freistaates Sachsen 2002). The gains in the 1990s can be attributed to a large degree to residential suburbanisation in the Dresden area. The data after 2000 reflect the phasing out of suburbanisation and also cyclical economic developments. The age selectivity of the migration streams poses particular problems: younger and well-educated people leave the region, and in particular younger women (BBSR 2013). A consequence of this development is an aging population structure. There are hopes that out-migration will decrease in the future, since the regional labour market and the job situation has improved considerably in recent years.

*Economic Structure and Key Sectors*

At the end of the 1980s it was obvious that the machinery, the industrial premises and the technical and transportation infrastructure of the region were dated and lagged behind the standards of Western industrial countries (Friedrich 2003). The Economic and Monetary Union and the introduction of a market economy that accompanied the reunification of Germany in 1990 consequently led to drastic changes in the economic structure of the region. Many enterprises were not competitive under market conditions and had to cease their operations. Job losses were especially high in the first years after transformation in the early 1990s. Employment in the textile industry, energy production and other branches of mass production decreased drastically.

A direct consequence of the closing down of many factories was a sharp rise in unemployment. Many workers left the region again during that period. The sharp drop in manufacturing employment was only partially compensated for by a growth of jobs in the service sector (Postlep 2004). New industries that settled in the region often had high productivity and employed only a fraction of the workers that the old industrial activities required (Friedrich 2003) with unemployment reaching a high of 72,700 in 2003 (April). The job market has however improved since this time. In 2011 only 35,400 persons were unemployed in the Oberlausitz Region with an unemployment rate of 12.4 per cent in 2011 (April) being the lowest in the region since the early 1990s (Agentur für Arbeit Bautzen 2011).

Although the majority of the open pit mines in Saxony have terminated their operation, the landscape and the economic structure of the north and north-east of the region are still influenced by the legacy of lignite mining. In contrast, the economy of the south and the south-west of the region is dominated by a mix of small and medium-sized enterprises (SMEs). Highly specialized firms, especially in the textile industry and machinery and engineering, can be found here. Friedrich (2003) sees a positive sign in the fact that industrial cores were renewed and stabilised in the region in spite of the economic turbulences. Today, about one-third of the workforce is employed in the manufacturing industry (Landkreis Görlitz: 33.5 per cent / Landkreis Bautzen: 35.8 per cent) (as of January 2012,

Statistisches Landesamt des Freistaates Sachsen 2012c). Key industrial sectors in the Oberlausitz are rail technology, mechanical engineering and metal processing, textile industry, plastic technology, food and beverage industry as well as energy production (MGO 2008). Each of the key sectors employs several thousand workers in the region. Rail technology is concentrated in a number of larger firms in the urban centres, including the largest employer, Bombardier Transportation GmbH, with manufacturing sites in Bautzen and Görlitz. Energy production is another key sector that is dominated by one company – namely Vattenfall. However, other key sectors are characterised more by a large number of SMEs (MGO 2008).

The Oberlausitz still has many traits of a lagging region. Indicators of structural weakness in the regional economy are very evident in the low rate of research and development activities, the dependency of many larger firms on external decision-making (branch plants), the low export rate and the low rate of new firm openings (ifo Institut für Wirtschaftsforschung Niederlassung Dresden 2005, based on Danielzyk and Zettwitz 2001).

## Global Challenges for the Oberlausitz Region

Two main challenges arise from the increased exposure of the Oberlausitz Region to international competition. The first challenge results from international competition for human capital: today all regions try to attract a talented and young workforce in order to get their own economies competitive and innovative. With the recent liberalization of the German labour market for Eastern European workers in May 2011, the Oberlausitz Region has new opportunities to attract a workforce directly from Polish and Czech neighbouring regions. However, fears and concerns of the local population regarding in-migration and the employment of a foreign workforce contrast with political strategies to enhance transnational mobility and cross-border exchange. A second major challenge arises from the emergence of global players in terms of the use of regional environmental capital. After years of state-driven destruction of environmental capital through lignite mining, the subsequent breakdown of the socialist exploitation and a process of restructuring and privatization, an 'energy giant' (*Vattenfall*) entered the Oberlausitz Region to continue exploitation. This global player is in a constant struggle with environmental protectors as well as regional land users that are negatively affected by mining activities. Simultaneously, new forms of land use can be observed: post-mining landscapes are being reclaimed for recreation and nature conservation as well as the growing renewable energy sector. These ways of using regional environmental capital offer new development potential but also create new conflicts.

### *Increased Mobility of People*

The migration of foreigners to and from the Oberlausitz Region can be seen as a phenomenon indicating integration into global migration flows. In terms of quantity, foreign population mobility is more important in the Oberlausitz Region than the international mobility of Germans. Altogether the region has profited from migration of foreigners during the last decade. The migration balance has been positive; with a net gain of approximately 4,500 between 2000 and 2011 (Statistisches Bundesamt 2013). Although there was a significant gain in foreigners through immigration from abroad, many of them left the Oberlausitz for other German regions after a short period. Consequently, the share of foreigners in the total population remained comparably small: 1.6 per cent in 2001 and 1.9 per cent in 2011 (Statistisches Landesamt des Freistaates Sachsen 2012c). There is no data available about the nationalities of the foreign population in the Oberlausitz Region. However, there is reliable data at the level of the Free State of Saxony (NUTS-1 level) which reflects the composition in the Oberlausitz. This data indicates that the 10 largest foreign groups in 2009 – according to their population size – originate from Vietnam, Russia, Ukraine, Poland, Turkey, China, Hungary, Czech Republic, Italy and India (Statistisches Bundesamt 2011). The list of these countries reflects the different immigration waves that were dominant in Eastern Germany. Throughout socialist times, Vietnamese people were contracted as an unskilled workforce and were sent back to their home country after apprenticeships – as part of the GDR's development aid (Kemper 2006). At the moment of reunification, Vietnamese people were the largest group of foreigners in the territory of the former GDR. Many stayed after reunification, and many of their children are naturalized Germans, but their population continues to shrink in statistical terms. Other Vietnamese people returned to Vietnam for retirement or for the newly dynamic economy (SMS 2009).

Russia and the Ukraine are countries from which a lot of resettlers (so-called *Spätaussiedler*) moved to Germany. These *Spätaussiedler* are ethnic Germans and their descendants were born in the territories of Eastern Europe, which had partially been under German control before the Second World War. They are granted a residence permit due to their German ethnicity upon arrival after 1993 into the reunified Germany. As they were also allowed to bring their non-German family members, most often Russians and Ukrainians and consequently their absolute numbers are quite important as well. Upon arrival in Germany, these resettlers are distributed to all counties according to Federal Law. While their papers are cleared after arrival in Germany, resettlers are not allowed to work and they have to be supported by German social security, which is paid for by the counties. In order to distribute this public effort in a fair way, each county has to take an equal share of resettlers as measured by the resident population and economic performance (Kemper 2006). Once the resettlers have received their official documents they are free to move to other regions. They often remain, however, in rural areas such as the Oberlausitz (Bucher et al. 2002).

Polish, Czech and Hungarian immigration is a clear indication of the Europeanization process. With the 2004 accession to the EU, people from these countries were free to settle in other EU countries (Kemper 2006). Even though Germany had legally limited access to its labour market until 2011, people from these countries immigrated in large groups to Saxony. The immigration of people from India and China to Saxony is linked to processes of globalization (Kemper 2006). In particular, university students and high-skilled workers in the IT sector have been attracted from these countries. The Oberlausitz Region, with the University of Applied Sciences Zittau-Görlitz and the *Internationales Hochschulinstitut Zittau* (IHI), is an attractive destination for students, especially from China (Sandner Le Gall and Mössner 2009). Turkish immigration has a long tradition in West Germany whereas in Eastern Germany it only began after reunification. In Saxony the presence of people from Turkey is not a consequence of guest worker immigration – as in West Germany – but it is often linked to Kurdish asylum seekers with Turkish nationality. Similar to resettlers from Russia, asylum seekers are distributed to all counties in Germany (Kemper 2006).

### *Attitudes of Resident International Migrants in Relation to the Oberlausitz Region*

During 2010 and 2011 qualitative interviews with resident immigrants were conducted in the Oberlausitz Region to analyse how foreigners assess their everyday life in this spatial context. The selection of interview partners was based on the principle of achieving very different biographical backgrounds. Thus, foreigners from 'distant' places (for example China, Vietnam, Mozambique) as well as 'close' places (for example Poland, Czech Republic, the Netherlands) were interviewed. They had arrived for jobs as well as for family motives; some had arrived during socialism, others more recently. As only 15 interviews were completed, they are illustrative rather than representative, but nonetheless they reveal a glimpse into the situation of foreigners in a rural area with a low share of foreign population. Compared to other regions in Europe the Oberlausitz Region is less attractive for international (working) migrants. However, some niche sectors of the regional labour market can be considered as being attractive in the demographic context and given the ageing of the population and the out-migration of young skilled people: the health and care service sectors are good examples. Despite lots of problems, the migrants in the Oberlausitz Region are confident about their employment situations and their lives. This is partly due to the fact that most of the interviewed migrants had arrived many years previous and felt attached to the region having developed circles of friends and acquaintances while also integrating into local public life (for example clubs and associations). They participate in volunteer and community work as well as regional policy-making.

While it can be concluded that migrants can be a very important human resource for the development of the Oberlausitz in providing, for example, an outsider's view on regional problems and opportunities for regional development, those interviewed did point to several problems. The local population for example is considered as rather narrow-minded and afraid of new cultural 'inputs' imported by foreigners. From the migrants' view this is a result of the GDR background and cultural homogeneity due to the restricted mobility of people during socialist times. Twenty years after reunification, the migrants assess this aloofness as a threat to the development of the region. As a consequence they try to become integrated by improving their language skills and their regional knowledge. This, however, is not easy particularly in the context of the Oberlausitz Region. The reasons for this include the fact that local public institutions and private service companies are not used to dealing with foreigners. Thus, service mentality towards the specific needs of international migrants is underdeveloped. There is also a poorly developed infrastructure (for example *ex ante* information systems, furnished housing, high-speed train connection, international airport and so on) which would facilitate arrival in the region.

### *Lack of Skilled Labour and Maintaining a Sufficient Labour Stock*

During the last two decades while the Oberlausitz Region has experienced complex processes of restructuring, it is still performing comparably poorly in economic terms. The regional labour market has been characterized by a surplus in skilled workforce on the one hand, and a high unemployment rate on the other. This imbalance pushed thousands of people to leave the region for more prosperous areas mainly in West Germany. In recent times the situation has changed: the regional labour market increasingly struggling with a lack of skilled labour (Arent and Nagl 2010). Until now, regional companies have been affected by post-socialist restructuring and increasing exposure to global competition. Nevertheless, they were able to profit from low wage levels and an oversupply of a skilled workforce. Since this time is coming to an end, regional companies more and more have problems in recruiting young educated and skilled staff to compensate the forthcoming wave of retirements. After years of out-migration of young and skilled people, the pool of potential workforce has shrunk dramatically. The few larger regional companies – often branch plants of multinational corporations – are able to adjust their wage levels and attract the remaining skilled people. Most regional companies however are small and medium sized, and consequently have problems in finding adequate personnel. This development is a serious threat to the future development of SMEs in the Oberlausitz Region. While the recently realized liberalization of the German labour market in May 2011 can offer new opportunities for the Oberlausitz to attract a skilled workforce from the neighbouring regions in Poland and the Czech Republic, it is questionable if a sufficient labour stock can be secured in order to maintain the productivity and competitiveness of the regional companies as wage levels are still below the national average.

### *Germany's Periphery: How to Become a Gateway Between Western and Eastern Europe*

Since the Second World War the Oberlausitz Region has been located in the periphery: first of the GDR and then of a reunified Germany. During GDR times the location at the Czech and Polish border did not offer many possibilities for exchange as the border transit was strictly limited. Encouraged by the geopolitical changes in Central and Eastern Europe, the borders became permeable and the situation changed significantly. This is reflected by intensified international migration as well as social, cultural and economic exchange. Since the Eastern enlargement of the EU in 2004 the Oberlausitz Region can no longer be viewed as a peripheral border location, but as one located at the interface of Eastern and Western European markets (Habermann 2007). Therefore, Saxony and in particular the tri-state border region may profit from this development as the region now has the opportunity to develop into a gateway between east and west, a pivotal location between Germany, Poland and the Czech Republic. However, based on the regional history of the twentieth century, mutual prejudices and distrust are still deeply rooted within peoples' minds in the border region and therefore the fostering of trust through the reduction of existing language barriers and the intensification of cultural exchange must be prioritized.

## Regional Responses to Global Challenges in the Oberlausitz Region

### *Improvement of Human Capital by Integrating International Migrants and Fostering Cross-border Mobility*

Endogenous development, according to Stockdale (2006), provides an opportunity for rural communities to become increasingly involved in their own futures. It is 'a self-centred and conservative process of growth that utilizes local experience and knowledge (van der Ploeg 1992) and assumes that the natural, human and cultural resources of an area hold the key to its sustainable development' (Francis and Henderson 1994, cited in Stockdale 2006, p. 356). A central aspect of endogenous development of rural areas is the need for appropriate human capital to be present. Continued out-migration and depopulation is not compatible with the endogenous development approach (Stockdale 2006). Human capital, however, can be redistributed over space through migration processes. Consequently, in-migration has the potential to introduce human resources (ibid.). According to Derounian (1998), migration to depopulated rural areas is 'potentially constituting something of a transfusion in the form of new blood, new ideas and fresh enthusiasm for locally based action' (cited in Stockdale 2006, p. 128). Other authors also note that the wealth of skills possessed by in-migration can provide an impetus to economic development (Dean et al. 1984, cited in Stockdale 2006). Thus, rural in-migration gives rise to the necessary human capital required for successful regional development.

### *Securing the Regional Workforce: Approaches in the Oberlausitz Region*

Several initiatives and associations which aim at the integration of international migrants into the regional labour market and local community and at the creation of a cross-border labour market have emerged in the Oberlausitz Region in recent years. Oberlausitz Neue Heimat e.V. (ONH) is an association established in 2002, which aims at integrating immigrants and deepening a democratic societal life. ONH defined ethnic German resettlers and their Russian family members as a specific target group for its activities. By organising and offering job-oriented measures (for example language training), the association helps resettlers to become economically autonomous from social security schemes. ONH, therefore, runs its own businesses which employ resettlers and other migrants. One of these businesses is the community centre (*Begegnungszentrum*) in Löbau, where resettlers and Germans gather and exchange ideas. The community centre is conceptualized as a multigenerational house (*Mehrgenerationenhaus*). Elderly people meet there in a sewing circle producing clothes for folk dancing groups. Young children and adults come together in theatre and dance projects. Furthermore, there is an international cooking group which offers catering to all kinds of events in the region. By integrating the resettlers into the association's activities, they are involved into the local public life instead of being a socially isolated group (Amadeu Antonio Stiftung 2013).

Another example is the regional cross-border partnership *Eures-TriRegio* which was initiated by Czech and Polish organizations after EU accession in 2004. After signing formal partnership agreements with German partners from the Oberlausitz in 2007, *Eures-TriRegio* works as a tri-national employment agency. Partners such as public labour offices, labour unions as well as employer groups from all three countries collaborate on the establishment of a harmonized and integrated cross-border labour market, where social standards and wages are respected. The engagement of labour offices and employers from the Oberlausitz is basically motivated by the search for a skilled labour force in the neighbouring Polish and Czech regions in order to respond to the growing scarcity of a skilled workforce (DGB Bezirk Sachsen 2013). Altogether seven *Eures-TriRegio* labour consultants have been hired to inform employers and employees from the Polish and Czech regions in their respective languages about their opportunities in the Oberlausitz labour market. Equally, job seekers from the Oberlausitz are informed in the German language about employment offers in Poland and the Czech Republic.

### *Making the Oberlausitz Region a Gateway Between Eastern and Western Europe*

Making use of the Oberlausitz' specific geographical location as a bridge between Western and Eastern parts of the EU can be seen as a future perspective for regional development. Several initiatives try to support this process. One such example is the *Internationales Begegnungszentrum St Marienthal* (IBZ) that was founded in 1992 by the local Cistercian convent (IBZ 2013). Today the organization exists in the form of a foundation under civil law (*Stiftung bürgerlichen Rechts*). The main goal of the IBZ is to bring people together, irrespective of their national origin or religion. The organization considers this as its contribution to peace, reconciliation of religions and worldviews and to mutual understanding of European nations and ethnic groups. The IBZ understands itself as an actor bridging east and the west as well as the present and future. Every year about 200 events take place in the IBZ dealing with different societal topics. Visitors from all over Germany and Europe contribute up to 19,000 overnight stays per year in IBZ facilities. Different projects aim at bringing together young people from Germany, Poland and the Czech Republic. Since 2005 young people from the border region's rural towns have been gathering in a tri-national youth parliament (*Trinationales Jugendparlament*) in the IBZ. Twice a year they learn about topics such as communal politics, EU politics, media competencies, public fundraising and environmental protection.

Since 1994 the *European School Network* has brought together 16 schools from ten European countries. The students of these schools jointly discuss Europe as a common home and make Europe an explicit issue for their school classes. The youth work outside of the school is also an important field of activities in the IBZ. Young people learn how to avoid racism and violence in integration seminars for ethnic German resettlers or in workshops on conflict resolution. International youth encounters are organized by the IBZ in order to engage young people with their foreign peers in discussion about current social issues from different national perspectives. These encounters help promote mutual understanding and tolerance. Young people from Saxony and the neighbouring Czech regions can participate in joint education programmes which qualify them for a professional future on both sides of the German-Czech border. Further projects focus on the general cross-border dialogue between all generations in the *Euroregion Neisse-Nisa-Nysa*. In addition, *Pontes* (lat: bridges), a specific project which has existed since April 2002, is located in the facilities of the IBZ. This tri-national education network aims at the development of the Euroregion towards a modern cross-border educational area within the context of a successful regional development. Its goal is to provide perspectives for a prosperous future for the region's population through the use of cultural and professional education and tri-national skills improvement as a key instrument (IBZ PONTES-Agentur 2012). *Pontes* defines its goals as: a) the development of euro-regional intercultural competency; b) the overcoming of demographic change by providing professional perspective for young people; and c) euro-regional education marketing across the German-Polish-Czech borders.

## Environmental Capital in the Oberlausitz Region Between Global Resource Scarcity and the Realignment Towards Sustainability

The Oberlausitz Region is characterized by diverse landscapes including the ecologically and scenically important area of the UNESCO Biosphere Reserve 'Upper Lausitz Heath and Pond Landscape' (*UNESCO-Biosphärenreservat Oberlausitzer Heide- und Teichlandschaft*). This valuable space is in close vicinity to areas that have been scarred or contaminated by mining and industry. Lignite mining and power-generation have been central to the regional economy for much of the last century. Although this sector has been subject to restructuring and modernisation after German reunification, it is still linked to a number of environmental and social problems. Simultaneously, new forms of land use can be observed such as the growing renewable energy sector for instance. While renewable energies can be found in nearly all parts of the region, mining activities are concentrated on its northern parts. Even today there are large deposits of electricity generation for the next decades. For this reason lignite deposits in the Oberlausitz are of great importance to ensure regional, and especially national, energy supply.

*Lignite Mining in Saxony and the Oberlausitz Region*

Saxony has a long tradition in lignite mining that goes back to the end of the nineteenth century when the demand for energy had risen constantly due to industrialisation. Even today the Free State of Saxony has large deposits that are concentrated in the Lausitz lignite fields and around the agglomerations of Leipzig and Halle (so-called Central German lignite fields). It is projected that these deposits will provide enough lignite for the current century. During the GDR, lignite was the main source of energy production and between 1945 and 1990 lignite mining was intensified to cover rising industrial demands for energy. Until the collapse of the state in 1989, the GDR was the world's largest producer of lignite, mining more than 300 million tons from 33 open pits per year (Murdoch et al. 2002). The exploitation of seams of lignite through strip mining involved massive changes for landscape and nature in the mining regions: entire villages disappeared, surface and groundwater were contaminated and the water table dropped by more than 30 metres in some places. Furthermore, lignite was not only detrimental to the land, but the process of converting it into energy produced large amounts of air pollution. The reunification of both German states laid the ground for a fundamental modernisation of the mining economy. Outdated and inefficient power plants were shut down and replaced by new and efficient ones. Large investments were made to restructure the whole energy sector in order to improve its efficiency and its environmental compatibility. Since 2002 the *Vattenfall Europe AG* (restructured and renamed to *Vattenfall GmbH* in 2012), a subsidiary enterprise of the state-owned Swedish energy company, *Vattenfall AB*, has operated two opencast mines (Nochten and Reichwalde) and the power plant Boxberg within the Oberlausitz Region. The opencast mine Nochten is located in the northeastern part of Saxony near the city of Weißwasser. Since 1973 the second Lusatian seam, and additionally since 2006, the first Lusatian seam have been exploited. Altogether, about 18 million tons of lignite is recovered at Nochten per year (Vattenfall GmbH 2012). As the first and second seams are located under massive soil layers, a lot of mine spoils have to be removed. The second mine, still under operation within the Oberlausitz Region, is the opencast mine Reichwalde. The exploitation of lignite in Reichwalde had been stopped in 1999 due to low efficiency, but in 2007 *Vattenfall* reopened the opencast mine and, at the beginning of 2011 extraction was restarted, and is expected to last for about 30 years.

The largest lignite power plant in Germany has been operated near the village of Boxberg since the 1970s, with an installed power-generation capacity of 3,520 MW. During the 1990s the old block-units were closed or rebuilt, but since 2000 a new block-unit with an installed power of 900 MW has been in operation. From 2007 to 2012 a further block-unit (675 MW) was built and put into operation. Both newly built block-units belong to the most modern and most efficient lignite power plants in the world. Compared to earlier times the economic importance of mining and power generation has decreased. *Vattenfall*, nevertheless, is still one of the largest enterprises in the region. In 2010 the business unit 'Lignite Mining & Generation' employed about 7,700 workers and 700 apprentices in the Lausitz lignite fields (in Saxony and Brandenburg) (Vattenfall Europe AG 2011). Moreover, a lot of ancillary industries and services are located in the mining area.

Although *Vattenfall* has undertaken a lot of efforts to reduce ecological impacts during the last decade, effects on the regional ecosystem are still enormous. Even today, landscape destruction and interventions into water and soil balances cause excessive environmental damage and evoke regional conflicts. Furthermore, settlements are still affected by mining activities and local inhabitants have to be resettled. While homeowners and municipalities receive compensation from *Vattenfall*, the payments cannot compensate for personal attachment to homes and neighbourhoods.

Interviews with stakeholders in the Oberlausitz Region show however that regional resistance to mining activities did not increase noticeably during the last decade. There is no broad protest movement in the region, a situation that can possibly be explained by different reasons:

- People in the Oberlausitz have grown up with mining activities, and feel traditionally attached to this industrial sector;
- During the last years *Vattenfall* developed effective instruments for civic participation (for example the establishment of open councils) and therefore affected inhabitants are informed and involved at an early stage of planning, or
- The mining sector is still the most important employer in the northern part of the Oberlausitz and a pillar of the regional economy.

### *The Renewable Energy Sector in Saxony and the Oberlausitz Region: How Does the Energy Transition Proceed?*

Alongside the exploitation of fossil fuels, alternative sources of power generation have become more and more important during recent years. Wind farms, solar parks and biogas power plants have been installed throughout Saxony to increase climate-friendly energy production, and allow greater independence from global energy markets. The development of the renewable energy sector (RES) in Saxony during the last decade has been very dynamic. From 2000 to 2010 the share of renewable energies in Saxon gross electricity consumption increased from 3.9 per cent to 15.3 per cent (SAENA 2012). The State Government of Saxony intends to increase the share of renewable energies to 24 per cent by 2020 (Freistaat Sachsen 2009). The wind power sector in the Oberlausitz Region developed very positively in the past, whereas the photovoltaic sector is booming at the moment due to decreasing installation costs and high subsidies. Despite this positive development, the contribution to regional added value of these power plants is questionable as these plants neither generate new jobs nor economic profit in the region. Some of these plants are large-scaled such as a solar park at the former airfield in Rothenburg/O.L. This solar park was constructed and is operated by the Munich Company *Gehrlicher Solar AG*. Almost all of such large-scale bioenergy projects in the region are initiated and financed by external corporate enterprises. Also, the technical equipment is not produced in the region, and maintenance work is done by external service companies. Consequently, gains from the production of renewable energy are not reinvested in the region nor do they act as further stimulation to regional development.

Local conflicts regarding the use of renewable energies, including environmental and social aspects, have emerged in recent years. Areas of conflict between renewable energies and environmental protection concerned mainly the protection of species, the ecological balance of the area and the conservation of landscapes. Social conflicts have been prompted by noise pollution (wind turbines and biogas plants), odour nuisance (biomass plants), shadow flicker (wind turbines) or visual despoilment (wind farms, solar parks, monoculture for the production of biomass). As a consequence, because almost all types of sustainable energies are affected by decreasing acceptance, the realization of new renewable energy projects is getting increasingly complex and time-consuming as more and more concerns have to be taken on board. Local resistance is reflected in an increasing number of citizens' initiatives that are trying to hinder the construction of renewable energy plants in their neighbourhoods (so-called NIMBY phenomenon – 'not in my backyard'). The following aspects can be seen as reasons for this development:

- Renewable energies are relatively new in the Oberlausitz and are perceived as a threat or danger by the local population;
- At the beginning of the bioenergy boom a lot of planning mistakes happened due to missing planning principles. For example, wind power plants and biogas plants were built near villages, impacting negatively on quality of life; and
- The effects of the bioenergy sector on the regional economy are assessed as being insignificant.

## Improvement of Environmental Capital by Promoting a Rural Eco-economy

In the global context of growing resource scarcity and the need to de-carbonise production and consumption systems, it is timely to reconsider the potential for combining the ecological with social and economic development in innovative ways (Kitchen and Marsden 2009). Kitchen and Marsden (ibid.) see 'the emergence of a new role for rural areas especially at the regional level, which can be encapsulated in the emergence of a new rural development paradigm' (p. 274). One central element of this new development paradigm concerns the need to take forward the more specific concept of the rural eco-economy – an economy that is integrated with its ecology, incorporating the idea of reciprocal production, exchange and benefits and embedded in its local physical and social territory. This section presents examples that can be important components of an emerging eco-economy within the Oberlausitz Region.

### *Development of the Lausitzer Seenland as a Tourist Region and Conservation Area*

Connected with the recultivation of post-mining landscapes, a new holiday region called *Lausitzer Seenland* is currently developing in the north-east of Saxony (*Oberlausitz*) and the south of Brandenburg (*Niederlausitz*) (see Figure 9.2). By flooding former mines, 23 artificial lakes will be created with a total water surface of about 350 $km^2$. This new holiday region will be the largest artificial lake-district in Europe, and the whole area around the *Lausitzer Seenland* is expected to become an attractive tourist region and holiday destination for national and international visitors (Zweckverband Lausitzer Seenland Sachsen 2013). The new lake-district will compete with traditional German holiday regions, for example the *Mecklenburger Seenplatte*. Hence, it will be important to develop unique features at each lake in order to improve its attractiveness for different tourist groups (for example adventurers, sportspeople, families with young children, elderly people). Today, only a few lakes have already reached the planned water level. In a few years, when the process of flooding is completed, ten lakes will be connected by navigable channels.

It is planned to create an attractive lake land that can be used for different types of sports-orientated leisure activities. In this context a lot of investment into local infrastructure is necessary. Beaches, marinas and lakeside promenades have to be built, as well as campsites, restaurants or centres for water-skiing and boating. Currently, initial projects have already been realized or are in the phase of realization. Several larger tourism projects such as the construction of hotel complexes and golf courses are planned to attract paying visitors. However, the realization of some of these large-scale projects has not started as of yet due to a lack of private investors.

The recultivation of former lignite mines and the restructuring of landscapes provide a lot of opportunities and development potential for the conservation of nature. Large, connected areas can be provided for nature protection, and a lot has been done to transform the mining areas into natural spaces again. Further, while this re-naturation has not been affected negatively by human activities due to the relatively small number of tourists coming to the region, this may change in future years as attraction to visitors grows.

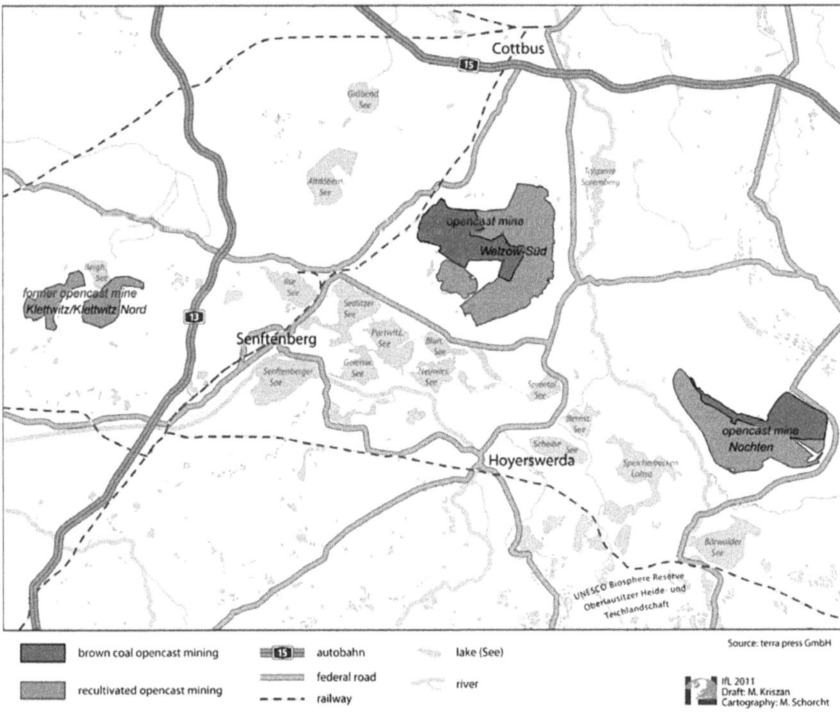


**Figure 9.2 Lausitzer Seenland**

## The Future Role of Renewable Energies

The main question for the future development of the regional RES sector is: What measures will help to raise the regional acceptance of renewable energies? In general it appears to be crucial that RES facilities are adjusted to regional potential (especially with regard to biomass-based projects) and also to landscape (especially regarding wind farms and solar parks). Large-scale facilities are linked to cost efficiency. However, they can have a negative impact on the natural balance and thus contribute to a negative perception of renewable energies among the population. Consequently, energy supply should be organized in a decentralized way. This opens possibilities for regional and local suppliers to enhance their engagement in the production of heat and electricity based on the utilization of renewable sources of energy. This approach can enhance the regional value-added and contribute to the economic development of the region. Moreover, heat and electricity supply could be organized 'close-to-the citizen', giving the local population the opportunity to participate directly in RES projects, for example through so-called citizen's power plants (*Bürgerkraftwerke*) and energy cooperatives (*Energiegenossenschaften*). These bottom-up approaches increase the acceptance of renewable energies and result in a better identification of the citizens with RES projects. Several positive examples already exist within the Oberlausitz Region, such as the citizens' co-operative Zittau-Görlitz (*Bürger-Energie Zittau-Görlitz* eG). The citizens' co-operative operates photovoltaic panels on public and private buildings in the county of Görlitz. As a first step of implementation, solar panels were installed on the roof of a regional credit bank in May 2010. Further photovoltaic panels were installed within just a short period of time in other villages of the county. The main aims of the co-operative are:

- production of environmentally friendly and non-polluting energy;
- decentralized production sites to avoid large losses of energy efficiency;
- gaining attractive profits for the shareholders of the co-operative;
- contracting of regional enterprises from handicrafts for the installation and maintenance of plants; and
- offering additional sources of income to house-owners (the leasing contracts between house-owners and the co-operative are closed for at least 20 years).

The citizens' co-operative was founded in September 2009, and by June 2010 it had recruited 48 members owning more than 300 co-operative shares. The co-operative can be seen as a new way for the local population to get engaged in, and to profit from, RES projects. Citizens get the opportunity to become shareholders and thus gain profits from the operation of renewable energy plants (Bürger-Energie Zittau-Görlitz eG 2012).

## Concluding Remarks – the Oberlausitz Region Reinvented

As a consequence of out-migration the regional economy has to struggle with an increasing lack of an educated and skilled workforce. In order to maintain their competitiveness, local enterprises seek to profit from the newly available international workforce from neighbouring countries as well as distant places. The local population, however, is still used to living according to patterns reflective of the 1990s homogeneous ethnic community. For that reason they remain reluctant towards in-migration of foreigners and emerging social diversity. Nevertheless, there are several endogenous initiatives, as well as regional strategies and policies, aimed at securing the regional workforce on the one hand, and improving the intercultural competencies of the local population on the other.

After the modernization and privatization of the mining sector in the Oberlausitz Region, the exploitation of lignite has been continued due to its importance for national energy supply in times of growing resource scarcity. However, the Oberlausitz has to develop a strategy for a post-lignite era. The establishment of a rural eco-economy can be an innovative way to sustain regional environmental capital on the one hand, and to create a new economic base on the other. The regional tourism sector – developing the *Lausitzer Seenland,* amongst others – as well as renewable energy production organized in a decentralized way can become main components of a future eco-economy.

In summary, it can be stated that the Oberlausitz has been affected by several contemporaneous processes which overlap in complex ways. The region tries to tackle the resulting challenges in a proactive way and, currently, the Oberlausitz is reinventing itself as it tries to find anchor points for successful future development.

## References

Agentur für Arbeit Bautzen (2011) *Arbeitsmarktreport April 2011*. Bautzen.

Amadeu Antonio Stiftung (2013) *Sächsischer Förderpreis für Demokratie. Löbau: Von der Selbsthilfe zum demokratischen Motor der Stadt* [Online: Amadeu Antonio Stiftung]. Available at: http://www.demokratiepreis-sachsen.de/die-preistraeger-der-vorjahre/preistraeger-2007/migranten-gestalten-demokratie/ [accessed: 20 January 2013].

Arent, S. and Nagl W. (2010) Ostdeutscher Fachkräftemangel bis 2030. *ifo Dresden berichtet*, 6/2010, 40–43.

Bastian, O., Porada. H., Roeder, M. and Syrbe R.-U. (eds) (2005) *Die Oberlausitzer Heide- und Teichlandschaft*. Leipzig. (= Landschaften in Deutschland 67).

Bundesinstitut für Bau-, Stadt- und Raumforschung (BBSR) (ed.) (2013) *Auf der Suche nach dem guten Leben*. BBSR-Analysen Kompakt, 4/2013. Bonn.

Bucher, H., Kocks, M. and Schlömer, C. (2002) Künftige internationale Wanderungen und die räumliche Inzidenz von Integrationsaufgaben. *Informationen zur Raumentwicklung*, 8, 415–30.

Bürger-Energie Zittau-Görlitz e.g. (2012) *Sonnenenergie: aus der Region – für die Region* [Online: Bürger-Energie Zittau-Görlitz e.g.]. Available at: http://www.buerger-energie-zittau-goerlitz.de/ [accessed: 20 January 2013].

Derounian, J.G. (1998) *Effective Working with Rural Communities*. Chichester: Packard Publishing Ltd.

DGB Bezirk Sachsen (2013) *EURES TriRegio – Informationen für Grenzgänger. Über uns* [Online: DGB Bezirk Sachsen]. Available at: http://www.eures-triregio.eu/index.php?option=com_content&view=article&id=60&Itemid=59&lang=de [accessed: 20 January 2013].

Freistaat Sachsen (ed.) (2009) *Sachsen 2020*. Wegweiser für unseren Freistaat. Strategisches Grundsatzpapier. Leipzig.

Freistaat Sachsen (2003) *Landesentwicklungsplan Sachsen*. Dresden.

Friedrich, K. and Schultz, A. (2007) Abwanderungsregion Mitteldeutschland. Demographischer Wandel im Fokus von Migration, Humankapitalverlust und Rückwanderung. *Geographische Rundschau*, 59 (6), 28–33.

Friedrich, W. (2003) Wirtschaftsgeschichte und Wirtschaftspolitik in der Lausitz, in Schröder, W. and S. Zundel (eds) *Regionale Entwicklung in der Lausitz*, pp. 14–55. Auf der Suche nach einem wirtschaftlichen Profil: Berlin.

Habermann, C. (2007) Die Wirtschaftsentwicklung in Sachsen. Erfolge und Probleme, in Geierhos, W. (ed.) *Regionalentwicklung der Oberlausitz*, pp. 16–23. Chancen und Perspektiven: Görlitz, Zittau.

ifo Institut für Wirtschaftsforschung Niederlassung Dresden (ed.) (2005) Die Region Oberlausitz-Niederschlesien – Innovationspotenziale und Perspektiven im Dreiländereck, *ifo dresden studien 38*. München.

Internationales Begegnungszentrum St. Marienthal (IBZ) (2013) *Über uns* [Online: IBZ]. Available at: http://ibz-marienthal.de/cms/front_content.php?idcat=4 [accessed: 20 January 2013].

Internationales Begegnungszentrum St. Marienthal (IBZ PONTES-Agentur) (2012) *Über PONTES* [Online: IBZ PONTES-Agentur]. Available at: http://www.pontes-pontes.eu/ueber-pontes.html [accessed: 20 January 2013].

Jain, A. and Schmithals J. (2009) Motive für die Wanderung von West- nach Ostdeutschland und Rückkehrtypen, in Cassens, I., Luy, M. and R. Scholz (eds) *Die Bevölkerung in Ost- und Westdeutschland – Demografische, gesellschaftliche und wirtschaftliche Entwicklungen seit der Wende*, pp. 313–33.Wiesbaden: VS Verlag.

Kemper, F.-J. (2006) Internationale Wanderungen und ausländische Bevölkerung in Deutschland. *Zeitschrift für Bevölkerungswissenschaft*, 31(3–4), 389–412.

Kitchen, L. and Marsden T. (2009) Creating Sustainable Rural Development through Stimulating the Eco-economy: Beyond the Eco-economic Paradox? *Sociologia Ruralis*, 49, 3, 273–94.

Kröhnert, S., van Olst, N. and Klingholz, R. (2005) *Deutschland 2020: Die demographische Zukunft der Nation*. Berlin.

Marketinggesellschaft Oberlausitz-Niederschlesien (MGO) (2008) *Wirtschaftsmagazin Oberlausitz 2007/2008*. Dresden.

Mudroch, A., Stottmeister, U., Kennedy, C. and Klapper H. (eds.) (2002) *Remediation of Abandoned Surface Coal Mining Sites*. Berlin.

Postlep, R.-D. (2004) Strukturwandel und Wirtschaftsentwicklung in Ostdeutschland – Wo stehen wir heute? in Stoll, M. (ed.) *Strukturwandel in Ostdeutschland und Westpolen*, pp. 128–32. ARL Arbeitsmaterial 311: Hannover.

Sächsische Energieagentur GmbH (SAENA) (2012) *Stand der erneuerbaren Energien in Sachsen* [Online: Sächsische Energieagentur GmbH]. Available at: http://www.sachsen-erneuerbar.de/Startseite/Erneuerbare_Energien_FS.html?PHPSESSID=d889e77e566350d2d5e428b93d58af31 [accessed: 20 January 2013].

Sächsisches Staatsministerium für Soziales (SMS) (2009) *Rahmenbedingungen von Migration und Integration im Freistaat Sachsen ab 1990* [Online: SMS]. Available at: http://www.soziales.sachsen.de/download/Download_Gesundheit/Migration_Sachsen.pdf [accessed: 20 January 2013].

Sandner Le Gall, V. and Mössner, S. (2009) Internationale Migration und Integration in Deutschland. *Geographie und Schule*, 31(177), 4–10.

Scheibe, B. (2011) Ergebnisse der 5. Regionalisierten Bevölkerungsprognose für den Freistaat Sachsen in ausgewählten Altersgruppen nach Kreisfreien Städten und Landkreisen bis 2025. *Statistik in Sachsen*, 17(1), 12–17.

Statistisches Bundesamt (2011) *GENESIS-Online Datenbank* [Online: Statistisches Bundesamt]. Available at: https://www-genesis.destatis.de/genesis/online/data;jsessionid=9B94BF33073AFCC3F44C403DF3E9721A.tomcat_GO_1_2?operation=abruftabelleBearbeiten&levelindex=2&levelid=1360325232342&auswahloperation=abruftabelleAuspraegungAuswaehlen&auswahlverzeichnis=ordnungsstruktur&auswahlziel=werteabruf&selectionname=12521-0005&auswahltext=%23SDLAND-14%23Z-31.12.2009%2C31.12.2008%2C31.12.2007%2C31.12.2006%2C31.12.2005%2C31.12.2004%2C31.12.2003%2C31.12.2002%2C31.12.2001&werteabruf=Werteabruf [accessed: 20 January 2013].

Statistisches Bundesamt (2013) *GENESIS-Online Datenbank* [Online: Statistisches Bundesamt]. Available at: https://www.regionalstatistik.de/genesis/online;jsessionid=0E20A22EFAD743E5CE128089243A5F35?operation=previous&levelindex=3&levelid=1360324039563&step=3 [accessed: 20 January 2013].

Statistisches Landesamt des Freistaates Sachsen (2012a) *Bevölkerungsbestand* [Online: Statistisches Landesamt des Freistaates Sachsen]. Available at: http://www.statistik.sachsen.de/html/426.htm [accessed: 20 January 2013].

Statistisches Landesamt des Freistaates Sachsen (2012b) *Statistischer Bericht. Bevölkerungsentwicklung im Freistaat Sachsen nach Gemeinden 2011* [Online: Statistisches Landesamt des Freistaates Sachsen]. Available at: http://www.statistik.sachsen.de/download/100_Berichte-A/A_I_2_hj2_11_SN.pdf [accessed: 20 January 2013].

Statistisches Landesamt des Freistaates Sachsen (2012c) *Regionaldaten Kreisstatistik Sachsen* [Online: Statistisches Landesamt des Freistaates Sachsen]. Available at: http://www.statistik.sachsen.de/appsl1/Kreistabelle/ [accessed: 20 January 2013].

Statistisches Landesamt des Freistaates Sachsen (ed.) (2002) *Sächsische Wanderungsanalyse. Ergebnisse einer Befragung 2002 bei ehemaligen sächsischen Bürgerinnen und Bürgern, die in der Zeit von Januar 2000 bis Juni 2001 in ein anderes Bundesland gezogen sind*, Sonderheft 3, Kamenz.

Stockdale, A. (2006) Migration: Pre-requisite for rural economic regeneration? *Journal of Rural Studies*, 22, 354–66.

UNESCO-Biosphärenreservat Oberlausitzer Heide- und Teichlandschaft (2013): *Das UNESCO Biosphärenreservat Oberlausitzer Heide- und Teichlandschaft* [Online: Biosphärenreservat Oberlausitzer Heide- und Teichlandschaft]. Available at: http://www.biosphaerenreservat-oberlausitz.de/allgemeines_zonierung [accessed: 20 January 2013].

Vattenfall Europe AG (2011) *Energie aus Braunkohle – Zahlen und Fakten 2010* [Online: Vattenfall Europe AG]. Available at: http://www.vattenfall.de/de/file/Zahlen-und-Fakten-2010-Energie_aus_Braunkohl.pdf_19343575.pdf [accessed: 20 January 2013].

Vattenfall GmbH (2012) *Tagebau Nochten/Reichwalde* [Online: Vattenfall GmbH]. Available at: http://www.vattenfall.de/de/tagebau-nochten-reichwalde.htm [accessed: 20 January 2013].

Zweckverband Lausitzer Seenland Sachsen (2013) *Herzlich Willkommen im Lausitzer Seenland* [Online: Zweckverband Lausitzer Seenland Sachsen]. Available at: http://www.lausitzerseenland.de/de/ [accessed: 20 January 2013].

# Chapter 10
# The Rural Regions of the Old-industrialized Saarland – Between Globalization and Regionalization

Birte Nienaber and Violetta Frys

## Introduction

Globalization increasingly shapes and challenges European regional development. Therefore the interlinkage between the global and the regional level are often of increasing importance. Elbe (2011) argues that regionalization is a specialization, while Benz and Fürst (2003) talk of spatial positioning and a concretion of globalization, regionalization and regional development together with globalization being two faces of one coin. Increasingly, internationalized property markets, large international enterprises and global actors represent new challenges for traditional environmental management. At the same time, however, globalization has reordered notions of 'core' and 'periphery' that have historically marginalized rural regions (Copus 2001, Dühr 2009, Eriksson 2008, Labrianidis 2004, McDonagh 2002, Wild and Jones 1993), and which now assign a whole new importance to rural areas.

Two main aspects of globalization are of interest for this chapter: international migration and regional learning in an era of globalization.

Due to the diverse impacts of globalization, the potential for individual mobility increases considerably. Faster and cheaper travel, especially the complex systems of mass air travel, favours international migration movements (Urry 2007, Adey 2006). Migration is often named as one of the main features of globalization (for example Karakayali 2011 and Ptak 2011). The new trends of international migration also include rural areas and bring new forms of migration into rural areas as well as internationalize the rural population. Cid Aguayo even argues that from this trend of international migration to rural areas evolves a 'global village' (Cid Aguayo 2008, p. 542). Flows in/to rural regions are not uni-directional, but are 'composed of movements into, out of, within and through rural places' (Milbourne 2007, p. 385). This plays an important role for the new trend of transnational migration. This form of migration does not only occur for example in big cities, it also affects people in rural areas. Transnational migration includes living and acting in more than one state and has impacts especially on larger metropolitan areas or on rural border areas (Nienaber and Kriszan 2013). These transmigrants live a 'special globalized life and build up their own social, economic and cultural spaces. The [ … ] cross-border social spaces that are being formed are serving to define, more and more, the social worlds of those involved. Such networks, once established, are themselves serving as facilitators of new waves of migration, of return, moving on and re-return. Indeed, the growing complexity of contemporary international migration pathways, nodes, moorings and way-stations- as well as their global reach, is without precedence' (Conway and Potter 2009, p. 1). The internationalization of home-ownership, labour migration and commuter movements in rural areas, such as those in Saarland, are also typical signs of a cultural globalization. Rapid exchange of ideas and cultural services can now more easily take place than ever before. The results of such cultural exchange are different and can be identified between standardization, cultural innovation and segmentation of specific local areas (Brock 2008).

Another focus of the research presented in this chapter is the arrangement of support for regional learning and innovation in rural Saarland with regard to globalization and regionalization. Learning and innovation are very important for rural regions because they develop capacities, knowledge and skills which are especially needed in economically poor border regions like Saarland. Grassroots initiatives in Saarland, therefore, help provide people in rural areas with new information, knowledge and skills as well as facilitate and support different stakeholders in learning to work together, to network and to reach a collective development aim.

Saarland is a federal state in Western Germany, with strong cross-border links to the neighbouring countries of Luxembourg and Northeastern France. The region is characterized by globalization processes, a very strong European cross-border integration process as well as internal regionalization processes. The region has a restructuring industrial economy, from coal and steel to a more diversified economy. It is an urbanized region and includes the city of Saarbrücken, which is also the state capital. Saarland is located in the so-called 'Greater Region', a term used to identify the transfrontier region Saar-Lor-Lux that contains the regions of Saarland, Lorraine, Luxembourg, Rhineland-Palatinate and Wallonia. The formerly used term 'Saar-Lor-Lux' referred primarily (since 1969) to the industrialized borderland of Saarland, Lorraine and Luxembourg. Furthermore, Saarland is integrated into other cross-border cooperations such as the 'Quattropole' (Saarbrücken, Trier, Luxembourg and Metz) and the cross-border metropolitan area. In these cooperations, the impact of globalization is to be found mainly in an urban context. Although Saarland is categorized as primarily an urban region by the OECD, there are rural parts of the Saarland that have gone through specific globalization and regionalization processes that cannot be compared directly with their urban counterparts. Consequently, this chapter focuses on two aspects of globalization that are obvious in parts of rural Saarland: international migration into rural northwestern Saarland; and capacity building, governance and knowledge systems as a globalization-driven regionalization process. By overcoming international borders, globalization is in this region very closely connected with the European integration process and Europeanization as well as regionalization in a transfrontier context.

## The Old-industrialized Federal State of Saarland as a Rural Case Study Region

Saarland has a total area of 2,570 square kilometres and is one the smallest of the 16 federal states of Germany, being only larger than the city-states of Hamburg, Bremen and Berlin (Statistisches Amt Saarland 2010a). Its unique history results from French and German influence: Saarland was economically affiliated with France from 1920 to 1935 and from 1947 to 1956. In the time between the formation of the Federal Republic of Germany in 1949 and the German reunification in 1990, Saarland was the only territory to rejoin Germany in 1957 after the Saar referendum. The state consists of six administrative districts: Merzig-Wadern, Sankt-Wendel, Saarlouis, Neunkirchen, Saarpfalz and Saarbrücken. Historically the region comprised a patchwork of industrial towns around coal mines and steelworks, with much of the population living and working in the same community. In the last half-century these industries have substantially declined. In 1957, there were nearly 64,000 people employed in 18 coal mines in Saarland, but by 2006 only 6,400 mining employees were left at one single mine (Dörrenbächer 2007). This process affected not only the coal mines, but also the steelworks.

Analysing the statistics and literature on Saarland, the following aspects of globalization are notable:

- The contribution of global competition, market forces and economic restructuring to deindustrialization, in particular the decline of the mining and steel industries and their associated sectors;
- The importance of foreign capital to Saarland industry, exemplified by the Ford car factory, and the integration of the region's major industrial employers into international trading networks, with the majority of their output exported;
- The increasing significance of cross-border ties in the 'Saar-Lor-Lux+-Greater Region', with cooperation in areas ranging from economic development to policing;
- Fluidity of movement across international borders with France and Luxembourg in both directions for work and residential relocation, including the settlement of Luxembourg citizens in the Perl community of Merzig-Wadern district; and
- The long-term presence of foreign migrant workers, especially from Southern Europe and Turkey, concentrated in Saarbrücken city and the smaller industrial and mining towns (Jones and Woods 2010).

The utilized agricultural area of Saarland (1,130 square kilometres) represents only 44 per cent of the total area. Forestry covers 33.4 per cent of the federal state, whereas settlement and traffic areas take up approximately 20 per cent. Cropland takes up more than half of the utilized agricultural area. Pasture land has a share of 35 per cent. The Natural Parks cover an area of 101,708 hectare (Statistische Ämter der Großregion 2008 and Brücher et al. 1982).

In 2010, little more than 1 million inhabitants lived in Saarland, equating to 396 inhabitants per square kilometre (Statistisches Amt Saarland 2011). The largest part of the population lives in the south, which includes the capital city of Saarbrücken and the traditional big industrial locations along the Saar River. Saarbrücken is a central place of high-order, which has a great appeal that goes well beyond the borders of Saarland. The state's urban agglomerations extend from Neunkirchen in the east to Völklingen in the west and from Saarbrücken to Saarlouis and Merzig in the northwest. The region of Saarbrücken forms a cross-border agglomeration area together with the Lorraine towns of Forbach, St Avold and Sarreguemines (Geppert 2007). The north of Saarland (Merzig-Wadern and Sankt-Wendel) and the southern part of Saarpfalz (Bliesgau) have a lower population density. They have a rural character and no major industrial hubs. The northwestern district of Merzig-Wadern has the lowest population density of the state, and was therefore chosen as the rural case study area for this migration research.

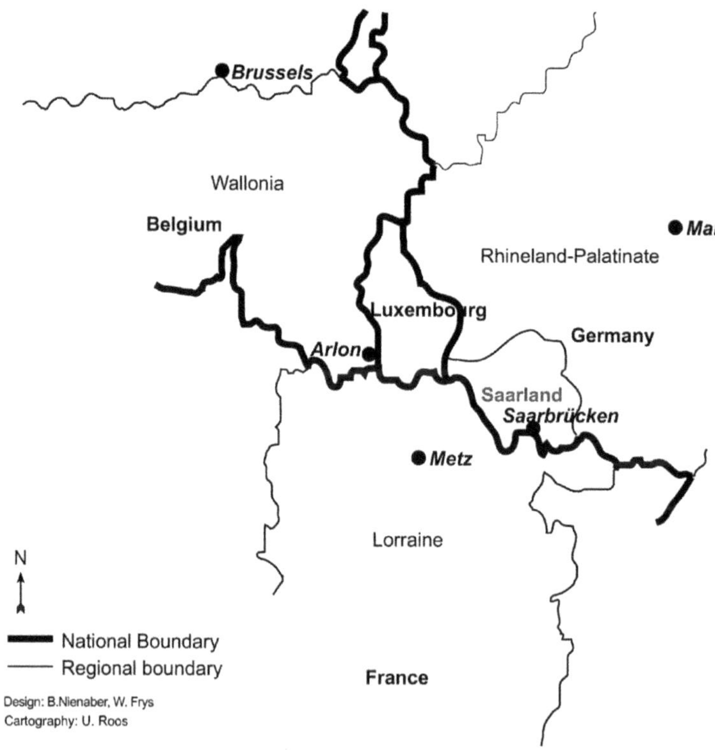


**Figure 10.1 The case study area and its neighbouring regions**
*Source*: author's graphic

**Migration into Saarland and the Neighbouring Countries as Part of a Globalization and Europeanization Process**

For decades, Saarland has been facing new and complex challenges. These can be thought of in the first instance as:

> The decline in the birth rate and the migration deficit are (…) the determining tendencies for the population development and they shape the process of ageing and the decline in the population of Saarland. (Ertl 2007, p. 21)

While the second important element is the:

> Declining population figures and an ageing society represent an enormous challenge for Saarland. For politics and economy, this means that they have to respond to these demographical processes and to form them. (Ertl 2007, p. 29)

Besides the high population density, there is a high birth deficit in Saarland: the birth deficit declines together with the population density, which means that the lower the population density, the lower the birth deficit is. In 2009, the birth deficit in Saarland varied between -545 in the district of St Wendel and -1743 in the regional district of Saarbrücken (Statistisches Amt Saarland 2010b). This is not surprising, since rural regions in Germany register higher birth rates than urban areas.

In comparison to internal migration, external migrations develop more dynamically. In addition to the economic situation of Saarland (especially with the coal and mining crisis and the decrease in employment opportunities), external migrations also depend on political situations, decisions and policies of the out-going and in-coming region. Thus, in-migration from abroad has been declining constantly due to the Asylum Procedure Act of 1993, although the 1990s also saw an increase in ethnic German migrants from Eastern European countries due to the fall of the Iron curtain and the declaration of these migrants as German citizens by Article 116 of the German 'Grundgesetz' (German Constitution). As such, the balance of international migration was in 2006 only slightly positive (+280). This was not enough to outweigh losses through domestic migration (total of -2066 persons in 2006). This trend is slightly inverted in 2010 (total +190 persons) (Ertl 2007, Grundgesetz für die Bundesrepublik Deutschland 2009, Statistisches Amt Saarland 2010d). In general, only the rural district of Merzig-Wadern registers a positive balance, resulting from the influx of Luxembourgers (+21 inhabitants in 2009) (Statistisches Amt Saarland 2010b).

A large amount of migrations in Saarland can be identified as cross-border migrations. These result from proximity to France and Luxembourg. There are two forms of cross-border migrations in Saarland: in-migration from Luxembourg (mostly employees who work in the highly developed south of Luxembourg) and out-migration to France (especially to the rural spaces of Lorraine). In a survey conducted in 2003, 67.8 per cent of Saarlanders and 43.6 per cent of Luxembourgers declared that they could imagine living in the neighbouring country (Cavet et al. 2006). An unusual aspect of the Greater Region are the so-called atypical migrants or commuters: while living in the neighbouring country (for example Luxembourgers in Saarland), these migrants still work in their home country (for example Luxembourg) and commute on a daily basis between the two nation states (IBA 2007a, Wille and Internationale Arbeitsmarktbeobachtungsstelle n.d.). These atypical cross-border commuters from Luxembourg have an impact on the rural areas of Saarland close to the border (district of Merzig-Wadern). An increasing number of atypical cross-border commuters from Luxembourg positively influence the balance of migration in the rurally characterized district of Merzig-Wadern.

In historical terms, the border between Saarland and Lorraine is both a place of German-French confrontation over centuries and of German-French reconciliation and European integration. The phenomenon of residence shifting was one indication for this recent past with the 'first traces of property and real estate acquisition by wealthy Saarlanders in nearby Lorraine could already be registered in the 1960s. The trend of cross-border leisure and tourism emerged, which intensified in the 1970s and, by acquiring more and more construction permits, Saarlanders have become increasingly resident in Lorraine in the following years' (translated according to Wille and Kuntz 2005, p. 51). In the 1980s, especially individuals from the cities of Saarbrücken and Saarlouis as well as middle-class citizens (mainly with children) from almost all communities of Saarland moved into the border-zone area of Lorraine. The properties, which were originally in demand, were the largest properties, located in the most favoured areas (for example good infrastructural connection to Saarland, high amenity value and low prices). Since the mid-1990s Saarlanders also bought agricultural land as these were properties awaiting a future designation as building land (Ramm 1999). More than 30 per cent of all 727 Moselle communities in France registered Germans (1998), of whom 90 per cent were Saarlanders (Wille and Kuntz 2005). However, this trend seems to have stopped since the early 2000s due to the increase in property prices in Lorraine (Wille and Internationale Arbeitsmarktbeobachtungsstelle n.d.). Concerning the motivation for the shifting of residence to France, 60 per cent named quality of life, like the rural and recreative ambience of the Moselle communities, as well as the French lifestyle (Wille and Kuntz 2005, Ministerium für Umwelt – Saarland and Préfécture de la Région Lorraine 2001). Wille and Internationale Arbeitsmarktbeobachtungsstelle (n.d.) claimed that the most important aspects for Saarlanders to move over to Lorraine are economic and financially driven ones.

There are also a large number of typical cross-border commuters who do not change their registered residence in the study region and whose movements shape Saarland and the whole Greater Region. Estimations by the European Council, on cross-border commuting in the Greater Region between the border-zones of the countries Luxembourg, Belgium, France and Germany, are more than ten times that for the EU-25 (EU-25: 0.4 per cent of employees, Greater Region: 4.2

per cent of employed people), illustrating the extraordinary importance of this phenomenon (Bläser and Wille 2009). While the number of the cross-border commuters employed in the Grand Duchy is constantly growing, more and more Luxembourgers move their residence across the borders to the neighbouring countries and commute to work. The Luxembourgian commuters consequently show preference for the German regions on the other side of, and close to, the border of their home country as their new residence. Between 1999 and 2006, the number of Luxembourgers with a residence in Germany has almost tripled. On the German side, they prefer the city of Trier and its rural hinterland and the northern Saarland (especially in the district of Merzig-Wadern and the municipality of Perl therein). In Belgium, the Luxembourgians are located especially in the border-zone areas of the region of Wallonia – in particular in the city of Arlon. On the French side of the border the Luxembourgish migrants settle down in northern Lorraine (for example in Thionville and Longwy). It is not only the Luxembourgish citizens however that prefer to settle in these border-zone areas of the Luxembourg neighbouring regions, but also citizens from other countries working in Luxembourg (Brosius and Carpentier 2010, IBA 2007b). This commuting and foreign homeowners' residential zone is within a corridor of 20 to 30 kilometres along the Luxembourgian border with good infrastructural connections. Decisive reasons for the residential shifting of Luxembourgers are, along with real estate prices in Luxembourg, high living costs, the wish to become homeowners as well as family, cultural and fiscal reasons. The downside of this development is the fact that the prices, especially for home ownership, real estate and construction ground, in areas bordering Luxembourg increase because of the demand of affluent customers of the Grand Duchy. This is also a disadvantage for the local, non-commuting population (Bläser and Wille 2009, Carpentier 2010, IBA 2007b). These developments evoke social, ecological and infrastructural problems. The rapidly increasing rental and real estate prices squeeze the local population and their successors from the residential market. Moreover, many border-zone communities develop into dormitory towns with less integration of new residents. The housing development areas mean new land use conflicts between the locals and the new residents. Furthermore, the increase of motorized individual transport between the residential areas and the working places causes increased traffic and consequent ecological problems (Bläser and Wille 2009, Carpentier and Gerber 2010).

Indeed it has also been argued by Wille and Kuntz (2005) that:

> closely connected to the phenomenon of the atypical German and Luxembourgian cross-border worker and foreign home owners are questions concerning Germans and French as well as German and Luxembourgian people living together. To meet the demand of attractive residential properties in Lorraine, municipal 'lotissements' (areas of new housing) were established in many communities. For local residents and newcomers, this results in Germans concentrating in one place and the Lorrainers staying among themselves in the original core of

> the settlement. This form of 'ghettoization' leads only partially to integration. Perceptions of coexistence between German newcomers and native Lorrainers are inconsistent, with 80 per cent of the Saarlanders questioned in one study considering themselves to be integrated whereas 86 per cent of the French mayors did not take this for granted, at least only partially. (p. 51).

The same situation is indicated for Perl. The municipality of Perl (district of Merzig-Wadern) has 7,583 inhabitants and is located in the border triangle. Perl borders the French département Moselle (Apach in the south) and Luxembourg (Schengen in the west). There are 101 persons per square kilometre, which is the lowest population density in Saarland (Statistisches Amt Saarland 2011) with 1,297 Luxembourgians living in the municipality of Perl in 2007 (Statistisches Amt Saarland 2010c). These Luxembourgians live particularly in the segregated and isolated areas of new housing.

## The Impact of Globalization in Saarland

The following analysis of the impacts of globalization is based on the research carried out in the EU FP7 project DERREG. The research methodology combines an intensive literature review, a statistical overview of secondary data of the federal state of Saarland as well as of the district and municipality level, 25 qualitative interviews with migrants, 27 expert and stakeholder interviews on migration, environment, planning, administration and regional learning throughout Saarland, as well as three expert workshops. All interviews and expert workshops took place during the DERREG project. The authors of this chapter carried out the interviews and workshops. The interviews were transcribed and analysed with a text analysis tool.

### *International Migration into Saarland*

Two main factors determine Saarland as a part of the globalized network of migration: the first is proximity to French and Luxembourgian borders at the 'heart' of the European integration process. Indeed the town of Schengen – where the free movement agreement ('Schengen agreement') was signed – is right in the middle of this border triangle. The second factor is the steel and coal mining history of Saarland, which provided jobs mainly for the so-called 'Gastarbeiter' (guest workers), who first came from Turkey, Italy, Spain and former Yugoslavia especially, contrasting current workers who mostly originate from Central and Eastern European countries; albeit guest workers now search for new job opportunities in the rural areas as the steel and coal mining industry decreases. The sample includes migrants from the Czech Republic, India, Turkey, Kazakhstan, Poland, China, Sri Lanka, Kyrgyzstan, Luxembourg, France and Russia. Indeed, because one of the main reasons for migrant workers to leave their home country and to move to the district of Merzig-Wadern is the improvement of their own economic situation which is often connected to a bad political or economic situation in their home country, there is often a feeling of being satisfied with the situation in the region of arrival:

> My motivation to come to Germany was the hope for a better future – simply to move to a country, where people have a higher standard of living than people in Poland at that time, in Communist Poland. Simply away from poverty and cruelty and to a better life. (Translated according to the interviewed migrant worker 6)

> We came here, because there has been war for some time in our city and that was exhausting for us. After a year, we emigrated to Germany. At that time, emigrating as fast as possible was the only opportunity we had. (Translated according to the interviewed migrant worker 10)

The second group of migrants, foreign homeowners, buy or build houses on property which is relatively cheap compared to prices in Luxembourg. Another reason for Luxembourgers to move to Saarland is the feeling of cultural foreignness. A special impact of globalization can be detected here: since Luxembourg hosts one of the main headquarters of the European Union, as well as being a globally important financial centre, a large number of migrants move into the country and to the Greater Region of Saar-Lor-Lux. The high percentage of foreigners (43.2 per cent of the population in 2011 (Statistiken der Großregion 2011) living in Luxembourg leads to this feeling of cultural foreignness of some people in their home country. The respondents expressed very positive views on their living situation in the case study region and were rather critical about their home country:

> The people here are really very nice. That's not the case in Luxembourg, I'm abroad here as well and for heaven's sake, I'm not a racist, but in our country [Luxembourg] that's too multicultural. It's getting out of hand. There were Cubans, my husband couldn't talk to them at all; my husband is Portuguese. There was a Belgian, there were Germans, there were Portuguese. (…) [In Luxembourg], everything is multicultural, that's too much. You can't [speak] your own language in your own country anymore. And here, I can still speak it. (Translated according to interviewed foreign homeowner 11)

This open border situation between Germany, France and Luxembourg is also a special advantage of a globalization or Europeanization process as it enables the people in the region to 'use the best' of each country. This is also a reason for migrants to move into this part of Saarland. Other interviewees, migrant workers as well as foreign homeowners, search for the amenities of rural landscapes, the calmness or the fresh air, and therefore choose the rural Saarland as their new residential place:

> Now we settled down here and I think it's very comfortable and, as I already said, the landscape is really beautiful. (…) Okay, it actually is a bit cheaper here, but that was not the essential motive why we came here. (…) Well, landscape, that you live here more on the countryside and not necessarily in the city centre. (…) 'Rural' was a criteria which was important for me. I have always lived in a rural environment; I am not a town person and so, that was a criterion. (Translated according to interviewed foreign homeowner 1)

The poor public transport and the absence of a larger city as well as the lack of jobs for foreigners and of ethnic association were mentioned as problems by some of the migrants:

> Foreigners should get more opportunities to find work more easily or to learn more at a language school. There should be more possibilities for foreigners. (Translated according to the interviewed migrant worker 3)

> In fact, I'm from North Rhine-Westphalia and what strikes me is that we had a lot of culture associations for all kinds of nations: Italians, the Spanish, they all had their own club. These associations also had a restaurant which was open for everybody and they organized a lot among themselves, celebrations, for example, but they did it for everybody. Something like that doesn't even exist here. (…) What I saw, but which isn't there anymore, was a Turkish meeting place in the pedestrian precinct. And that was really the only thing I've seen so far in this area. And only Turkish people actually went there, not one single German. This custom of inviting each other does not take place. (Translated according to the interviewed migrant worker 9)

Some of the respondents, especially foreign homeowners, did not make a big effort to become integrated into Germany as a result of their professional integration and social rootedness in Luxembourg. Furthermore, the Luxembourgian migrants did not intend to change their jobs because they felt more secure with the more favourable tax policy and the well-balanced health care system of their home country. They live in both countries, benefiting from 'the best' of each. Processes and impacts of globalization in rural Saarland are therefore evident in the range of home countries of immigrants, as well as in the transnational way of life of most of the respondents.

## Regional Learning in a Globalized Context in Saarland

Strategies and support for 'regional learning' are of great importance in Saarland. These strengthen social and human capacity for regional development, and can be positioned as part of a regionalization response to globalization. Saarland is one of the NUTS 2 regions of the EU and therefore participates in EU-funded programmes, instruments and initiatives (for example ESF). Moreover, it has access to funds of the national and federal state government (for example Weiterbildungsdatenbank Saar (database for further education) or the project 'LernortIch'). The most important and effective form of support and facilitation from public administration appears to be financial support by means of formal subvention contracts like LEADER or ILE (integrated rural development concepts). One example is the initiative KuLanI (Cultural Landscape Initiative St Wendeler Land) which is a LEADER local action group, established in the 1990s as a grassroots initiative and transformed into a LAG (local action group) in 2003. The main focus of KuLanI is preserving and further developing the cultural landscape of 'St Wendeler Land' (an area in north-east Saarland, bordering Rhineland-Palatinate). There are three spheres of activity: raising awareness (local schools and school camp 'BiberBurgBerschweiler'), organizing a local commodity market ('Lokalwarenmarkt St Wendeler Land') and supporting tourism ('St Wendeler Land – Steinreich'). These focal points ensure the support and facilitation of the grassroots development initiative with grants and supports from the EU, federal ministries' and Saarland ministries' policies. At this point, it is possible to detect a future internationalization: KuLanI is building up contacts to French, Polish, Lithuanian and Luxembourgian grassroots initiatives for knowledge exchange. Other examples are Biosphere Bliesgau Association (Biosphärenverein Bliesgau e.V.), Integrated Rural Development Concept Region Illtal (ILEK – Integriertes ländliches Entwicklungskonzept) or the Saar Countrywomen's State Association (Landesverband SaarLandFrauen e.V.).

Even if some networks have an artificial character, the networking in the region facilitates knowledge transfer, the dissemination of support and sustainable development. Most of the networks are judged to be sustainable after being established. Analysing the support and knowledge infrastructure in Saarland, it is possible to identify opportunities and risks for integration into global networks. Public information and advisory services were evaluated as partly complex and difficult to use. Sometimes, there are different contact persons in different authorities that all have to be included as each authority has different information strategies and different regulations that have to be followed. The easy accessibility and availability of new information, however, are the most important advantages of this support form in rural Saarland. The most crucial and sustainable support for grassroots development initiatives in the future is the establishment of networks mediated by public authority. The grassroots development initiatives appreciate the opinions of external experts, who they mostly know personally, and obtain the latest knowledge available in this way. Also, scientific studies concerning grassroots development initiatives are seen as professional and efficient. The high costs of scientific research and resources, however, do not allow the grassroots development initiatives to use this form of support very often. Similar aspects apply to various stakeholders from other knowledge institutions of public authorities. No costs however are involved if consultants at federal state level, for example at the ministries, provide their support. Those consultants are paid by the state within the scope of their position.

Within these chosen initiatives a process of learning occurs from each other. The members' learning process includes not only the work within their own initiative but also cooperation with partners of other projects, whether established or potential ones. For this reason, it is also a question of a learning process in which the participants can broaden their horizons and where they can casually proof which new cooperation partners are good to work with (for example other communes). Apart from this fact the employees and members join different advanced training courses, meetings, conventions or skill-enhancement workshops. One example is the strategy meeting which is organized every six months by the regional manager (cooperation partner of the initiative). This assistance is project-oriented and based on specialist counselling and a research project. There are also training, workshops and meetings with experts for members of the initiatives concerning certain topics (for example, medical science, insurances, communication or legal matters). Afterwards, the acquired knowledge is passed on to other members:

> Nowadays there is this trend, so everything is anonymous. Information will be exchanged via email, SMS, data transfer, on electronic ways. A counter-pole will be established there [by the Saar Countrywomen's State Association]. The members of the initiative maintain a personal contact and this personal level is another way of communication. So we go in contrary to the trend in society to do everything in the chat room, on the internet and say: local exchanges to each other strengthen the social network that serves everybody, serves the organization because this is our elixir. This is basically the database; the personal contact is our database. Of course, we have a modern office communication, but it is not possible without the personal connection and I think that's why we are sitting here together on a personal level to get to know. And it binds and that's the goal that this bond is estimated, because it supports each other. So, the members support the organization that can use this support to perform their work (…) and the organization as a service provider gives something back to the members, (…) e.g. by building a network. (Translated according to an interview partner)

Furthermore, according to the initiative, support exists from their own ranks, allowing information and knowledge to be exchanged at different levels (for example local or national level). Knowledge institutes often offer an advisory support free of charge. Thus, the selection of topics is carried out in a system-oriented, project-oriented and territorial-oriented way.

Albeit there can certainly be a demand for additional support, for example by charitable foundations. A definite know-how is the condition for a successful collaboration, as well as the consideration of further cooperative domains with other organizations or institutions. One important support refers to the acquisition of new members who would join just for the project itself. Moreover, the acceptance of the population is generally of great importance. Depending on the topic, the initiatives foster regional identity which can only be seriously recognized through acceptance. To receive this support the initiatives require the encouragement of public institutions in order to appeal to potential cooperation partners and members through public campaigns.

## Conclusions

Globalization has different impacts across rural Saarland. Saarland, with its coal mining and steel history along the river Saar in the centre of the region, and the rural regions at its periphery, promotes different processes. In this chapter, the focus was on the rural regions at the border of France and Luxembourg on one side, and the northeastern rural areas bordering Rhineland-Palatinate on the other. The research has shown particularly that migration is not only a phenomenon of (global) cities, but can be a significant process in rural areas when the framework for migrants – politically, historically, economic, fiscal, social, cultural and so on – is given. Saarland has historically shifted several times between France and Germany, and consequently the region has always been connected with migration flows as well as people from different countries living next to each other – even in rural areas. Within this, a very important group of international migrants are the so-called 'Gastarbeiter'(guest worker) coming mainly in the 1960s and early 1970s from South European countries, Turkey and other countries and new forms of economic workers coming since the 1990s from Central and Eastern European countries. Another very important aspect regarding migration is the proximity and good infrastructural connections to economically thriving Luxembourg as well as proximity to the village of Schengen – the so-called 'heart of European integration'. These are the main aspects that constitute the breeding ground for migration as the main part of globalization processes in the district of Merzig-Wadern in Saarland. The globalization process leads to increases of multiculturalism and transnationalism due to the rise in the number of international migrants. The cross-border migrants particularly, but also many of the labour migrants, build up transnational spaces and networks integrating their former and their new residential countries into the live concepts. The increase of networks and connections to other parts of the world based on international migration is a second implication of globalization in this region. By overcoming international borders, globalization in this region is very closely connected with the European integration process and Europeanization as well as regionalization in a transfrontier context as could be seen in the analysis of our research as well as in the case of 'Schengen'. Besides these more positive challenges, the rural region of Merzig-Wadern in Saarland also has to overcome several difficulties due to the different forms of increasing international migration: for example the difficulties of integration as the Luxembourgians live in their own 'luxurious ghettos' and the problems of language acquisition for all migrant groups as the infrastructure of language schools are quite poor in rural areas. Another interesting aspect of this research was the reason for moving from Luxembourg to Germany due to perceptions of cultural foreignness in the home country. This can also lead to integration problems as migrants become foreigners in the receiving country as well.

So, for further research studies, it seems to be important to have a closer look at rural areas not only as donors of migrants, but also as recipients of migrants – even international ones. The second process, stated above, that shows a somehow more anti-globalized, regionalized process that seems to be strengthened due to globalization, include the learning and support strategies (except the support and guidelines of the European Union). In this case, a focus on the regional level, on regional stakeholders and regional or national support can be informative. The global faces of the coin, in keeping with Benz and Fürst (2003), are the influences of internationally learning cooperations as well as external support, policies and guidelines that form the framework in which an initiative like KuLanI, for example, can arrange its regional networks and regional development. These are just some of the aspects of globalization in Saarland; there are many others (for example the environmental management of UNESCO biosphere reserve Bliesgau or the connection of enterprises). Further insight on the initial aspects deduced from the literature and statistical review can be found in the case study contextual report (Jones and Woods 2010) as well as in articles on the environmental capital of the biosphere reserve (Frys and Nienaber 2011, 2013).

## References

Adey, P. (2006) 'If mobility is everything then it is nothing: towards a relational politics of (im)mobilities', *Mobilities* 1, 75–94.

Benz, A. and Fürst D. (2003) Region – 'Regional Governance' – Regionalentwicklung, in Adamaschek, B. and Pröhl M. (eds) *Regionen erfolgreich steuern. Regional Governance – von der kommunalen zur regionalen Strategie*. Gütersloh, 11–66. Available at: http://www.bertelsmann-stiftung.de/cps/rde/xbcr/SID-OEC8EF64–65BA54A4/bst/xcms_bst_dms_15213_15214_2.pdf [accessed: 16 January 2013].

Bläser, R. and Wille C. (2009) Grenzgänger im Großherzogtum Luxemburg, *Geographische Rundschau*, 61 (1), 36–42.

Brock, D. (2008) *Globalisierung. Wirtschaft – Politik – Kultur – Gesellschaft*. Wiesbaden: VS Verlag für Sozialwissenschaften.

Brosius, J. and Carpentier S. (2010) Grenzüberschreitende Wohnmobilität von in Luxemburg ansässigen Erwerbstätigen, in Carpentier, S. (ed.) *Die grenzüberschreitende Wohnmobilität zwischen Luxemburg und seinen Nachbarregionen. (Forum Europa, Vol. 6)*, pp. 15–36. Luxembourg: Édition Saint-Paul.

Brücher, W., Quasten, H. and Reitel F. (eds) (1982) *Pilotstudie zu einem Saar-Lor-Lux-Atlas. (Schriftenreihe der Regionalkommission Saarland-Lothringen-Luxemburg-Rheinland-Pfalz, Vol. 8)*. Saarbrücken, Metz, Luxembourg, Trier: Regionalkommission Saarland-Lothringen-Luxemburg-Rheinland-Pfalz.

Carpentier, S. (2010) Gründe für den Umzug auf die andere Seite der Grenze, in Carpentier, S. (ed.) *Die grenzüberschreitende Wohnmobilität zwischen Luxemburg und seinen Nachbarregionen. (Forum Europa, Vol. 6)*, pp. 37–62. Luxembourg: Édition Saint-Paul.

Carpentier, S. and Gerber P. (2010) Welche Konsequenzen hat die grenzüberschreitende Wohnmobilität auf die täglichen Fahrten und die Aktivitätsräume, in Carpentier, S. (ed.) *Die grenzüberschreitende Wohnmobilität zwischen Luxemburg und seinen Nachbarregionen. (Forum Europa, Vol. 6)*, pp. 87–113. Luxembourg: Édition Saint-Paul.

Cavet, M., Fehlen, F. and Gengler C. (2006) *Leben in der Großregion. Studie der grenzüberschreitenden Gewohnheiten in den inneren Grenzräumen der Großregion SaarLorLux/Rheinland-Pfalz/Wallonien. (Forum Europa, Vol. 2)*. Luxembourg: Édition Saint Paul.

Cid Aguayo, B.E. (2008) Global Villages and Rural Cosmopolitanism: Exploring Global Ruralities. *Globalizations*, 5 (4), 541–54.

Conway, D. and Potter, R.B. (2009) Return of the next Generations: Transnational Migration and Development in the 21st Century, in Conway, D. and Potter, R.B. (eds) *Return Migration of the Next Generations. 21st Century Transnational Mobility*, pp. 1–16. Farnham.

Copus, A. (2001) From Core-periphery to Polycentric Development: Concepts of Spatial and Aspatial Peripherality, *European Planning Studies*, 9(4), 539–52.

Dörrenbächer, P. (2007) 50 Jahre Saarland – 50 Jahre Kohlekrise. Die Entwicklung des saarländischen Steinkohlenbergbaus, in Dörrenbächer, P., Kühne, O. and Wagner J.M. (eds) *50 Jahre Saarland im Wandel*, pp. 101–13. Saarbrücken: Institut für Landeskunde im Saarland.

Dühr, S. (2009) Regional development and regional policy in the Netherlands: are there peripheral regions?, *Raumforschung und Raumordnung*, 67(4), 300–307.

Elbe, J. (2011) *Die Wirksamkeit von Sozialkapital in der Regionalentwicklung. Erprobung eines anwendungsorientierten Analysekonzepts am Beispiel des Modellvorhabens 'Bioenergie-Regionen'*. Aachen: Shaker.

Eriksson, M. (2008) (Re)producing a 'peripheral' region. Northern Sweden in the news, *Geografiska Annaler, Series B*, 90(4), 369–88.

Ertl, D. (2007) Bevölkerungsentwicklung 2006: Stärkster Bevölkerungsrückgang seit 1978, *Statistisches Quartalsheft Saarland III*, 2007, 21–9.

Frys, W. and Nienaber, B. (2011) Protected Areas and Regional Development: Conflicts and Opportunities – Presented on the Example of the UNESCO Biosphere Reserve Bliesgau. *European Countryside*, 3, 208–26. DOI: 10.2478/v10091-012-0005-8.

Frys, W. and Nienaber, B. (2013) Conflicts and Opportunities for Regional Development arising from the Designation of Protected Areas – the example UNESCO Biosphere Reserve Bliesgau, in Kriszan, M. and Burdack, J. (eds) *Use or Protect – Environmental Capital and Regional Development in the Rural Areas of Europe. Forum ifl*, Heft 20, pp. 14–35. Leipzig: Leibniz-Institut für Länderkunde.

Geppert, J. (2007) *Regionalprofil Saar-Lor-Lux-Rheinland-Pfalz*. Großrosseln: EURES Transfrontaliers Saar-Lor-Lux-Rheinland-Pfalz.

*Grundgesetz für die Bundesrepublik Deutschland. Stand vom 15. Juli 2009.* Baden-Baden: Nomos.

Interregionale Arbeitsmarktbeobachtungsstelle (IBA) (2007a) *Bericht zur wirtschaftlichen und sozialen Lage der Großregion. Bericht der Interregionalen Arbeitsmarktbeobachtungsstelle für den Wirtschafts- und Sozialausschuss der Großregion*. Saarbrücken: Interregionale Arbeitsmarktbeobachtungsstelle.

Interregionale Arbeitsmarktbeobachtungsstelle (IBA) (2007b) *Die Arbeitsmarktsituation in der Großregion. 5. Bericht. (Bericht im Rahmen des 10. Gipfels der Großregion)*. Saarbrücken: Interregionale Arbeitsmarktbeobachtungsstelle.

Jones, L. and Woods, M. (with input from B. Nienaber and W. Frys) (2010) *Case Study Contextual Report 5: Saarland (DERREG – Developing Europe's Rural Regions in the Era of Globalization), Aberystwyth*, Available at: http://www.derreg.eu/system/files/Saarland%20WP5%20contextual%20report.pdf [accessed: 3 September 2011].

Karakayali, S. (2011) Migration und Flucht, in Niederberger, A. and Schink, P. (eds): *Globalisierung. Ein interdisziplinäres Handbuch*, pp. 180–187. Stuttgart/Weimar: Verlag J.B. Metzeler.

Labrianidis, L. (2004) *The Future of Europe`s Rural Peripheries*. Aldershot: Ashgate.

McDonagh, J. (ed.) (2002) *Economy, Society and Peripherality, Experiences from the West of Ireland*. Dublin: Arlen Academic.

Milbourne, P. (2007) Re-populating rural studies: Migrations, movements and mobilities, *Journal of Rural Studies*, 23, 381–6.

Ministerium für Umwelt – Saarland and Préfécture de la Région Lorraine (eds) (2001) *Grenzüberschreitende Raumordnungsstudie Saarbrücken-Moselle-Est. (Schriftenreihe der Regionalkommission Saarland-Lothringen-Luxemburg-Trier/Westpfalz. Arbeitsgruppe Raumordnung, Vol. 12.* Saarbrücken, Metz: Regionalkommission Saarland-Lothringen-Luxemburg-Trier/Westpfalz.

Nienaber, B. and Kriszan, A. (2013): Entgrenzte Lebenswelten: Wohn- und Arbeitsmigration als Ausdruck transnationaler Lebensentwürfe im deutsch-luxemburgischen und deutsch-polnischen Grenzraum, *Raumforschung und Raumordnung*, 71 (3), 221–32. DOI: 10.1007/s13147–013–0230–2.

Ptak, R. (2011) Ökonomie, in Niederberger, A. and Schink, P. (eds) *Globalisierung. Ein interdisziplinäres Handbuch*, pp. 9–23. Stuttgart/Weimar: Verlag J.B. Metzeler.

Ramm, M. (1999) Saarländer im grenznahen Lothringen. 'Invasion' oder Integration?, *Geographische Rundschau*, 51(2), 110–115.

Statistiken der Großregion (2011) *Anteil der ausländischen Bevölkerung*. Available at: http://www.grande-region.lu/eportal/pages/StatTemplate.aspx?view=stat&id=471 [accessed: 3 September 2011].

Statistische Ämter der Großregion (ed.) (2008) *Saar-Lor-Lux-Rheinland-Pfalz-Wallonie – Statistiques en bref 2006, Statistische Kurzinformationen 2008*. Saarbrücken, Nancy, Luxemburg, Bad Ems, Namur: Landesamt für Zentrale Dienste.
Statistisches Amt Saarland (ed.) (2010a) *Fläche, Bevölkerung in den Gemeinden am 30.06.2010 nach Geschlecht, Einwohner je km² und Anteil an der Gesamtbevölkerung*. Available at: http://www.saarland.de/dokumente/thema statistik/staa_FB300610.pdf [accessed: 8 November 2010].
Statistisches Amt Saarland (ed.) (2010b) *Statistik: Grafiken, Tabellen und Artikel*. Available at: http://www.saarland.de/6772.htm [accessed: 4 November 2010].
Statistisches Amt Saarland (ed.) (2010c) *Ausländische Bevölkerung in den Kreisen des Saarlandes am 31.12.2007 nach EU-Staatsangehörigkeiten*. Unpublished documents.
Statistisches Amt Saarland (ed.) (2010d) Wanderungen 2010. *Statistische Berichte AIII1, j 2010*. Available at: http://www.saarland.de/dokumente/thema_statistik/STALA_BER_AIII1-J-10.pdf [accessed: 15 January 2013].
Statistisches Amt Saarland (ed.) (2011) *Tabellen und Grafiken aus dem Bereich 'Gebiet und Bevölkerung'*. Available at: http://www.saarland.de/dokumente/thema_statistik/staa_FB311210.pdf [accessed: 15 August 2011].
Urry, J. (2007) *Mobilities*. Cambridge and Malden: John Wiley & Sons.
Wild, T. and Jones P. (1993) From peripherality to new centrality? Transformation of Germany's Zonenrandgebiet, *Geography*, 78, 281–94.
Wille, C. and Internationale Arbeitsmarktbeobachtungsstelle (n.d.) Atypische Grenzgänger nach Luxemburg. *GR-Atlas*. Available at: http://geo.uni.lu/joomla/index.php?option=com_content&task=view&id=1820&Itemid=660 [accessed: 15 January 2013].
Wille, C. and Kuntz L. (2005) Im eigenen Land wohnen, beim Nachbarn arbeiten? Komplexität und Vielschichtigkeit des Grenzgängerwesens in der Großregion am Beispiel deutscher Grenzgänger und von Grenzgängern mit Wohnsitz in Deutschland, in EURES and OIE (eds) *Dimensions socio-économiques de la mobilité transfrontalière. Actes du séminaire transfrontalier EURES-OIE*, pp. 47–55. Luxembourg: EURES.

## Chapter 11

# Globalization and Sustainable Development in South Moravia

Milada Šťastná

### Introduction

The South Moravian Region (Jihomoravský kraj in Czech, and previously known as Brněnský kraj), occupies the southeastern corner of the Czech Republic, bordering Austria and Slovakia. It is a largely rural region, but contains the sizeable city of Brno as its capital. These geographical characteristics are significant in shaping the region's experience of globalization, with connections to international networks and processes commonly taking the form of shifting cross-border linkages, or being mediated through Brno. At the same time, globalization has touched on all aspects of the rural economy and society in South Moravia, bringing both challenges and opportunities for sectors such as wine-making and tourism, both of which are intimately tied to the region's distinctive cultural and environmental resources. This chapter describes the socio-economic and environmental characteristics of the South Moravia Region, examines the impact of globalization and highlights some of the initiatives in the region that have responded to the challenges and opportunities of globalization.

As with most of Central and Eastern Europe, the approach to economic development in the South Moravian Region has undergone a complete change in direction since the fall of communism in 1990. The previous system of centrally planned specialized industrial development has been replaced not only by a market-driven approach, but also by an emphasis on bottom-up development and entrepreneurship. Economic liberalization has encouraged foreign direct investment and this has been an important feature of South Moravia's engagement with the global economy, as noted below, but its reach has been limited, especially in more rural parts of the region, where greater emphasis has been placed instead on local resources and capacities. The concept of neo-endogenous rural development (Ray 1998, 1999, 2006) can be drawn on to characterize this latter approach, involving the mobilization of local actors and resources, but also the selling of the region to extra-local consumers, and the injection of external capital, often in the form of EU structural funds. The strategy has focused on the exploitation of the region's 'culture economy', described by Ray (1998, p. 3) as an attempt 'to localize economic control – to (re)valorize place through its cultural identity'.However, as van der Ploeg and Marsden (2008) and others have documented, the successful implementation of neo-endogenous rural development requires the effective interaction of local resources, innovation, social capital and governance frameworks. It is the conditions for these dynamics in the South Moravian Region that are examined in this chapter, with examples of good practice highlighted.

## Characteristics of the Region

As noted above, South Moravia occupies a peripheral location in the Czech Republic, with international borders with Austria and Slovakia that have changed their geopolitical significance over recent decades (the border with Austria was part of the closed 'Iron Curtain' until 1989, whilst the frontier with Slovakia was an internal boundary until the dissolution of Czechoslovakia in 1993). At the same time, the region lies in the geographical centre of the new Europe, closer to Bratislava and Vienna than to the Czech capital, Prague. Yet, both South Moravia and its neighbouring regions are relatively marginal within the Czech Republic and Slovakia (Vaíshar et al. 2011).

The region has a largely rural landscape of fertile lowlands in the south and west, and partly forested uplands and karsts landscapes in the north and east. The southern border is partly formed by the Morava and Dyje rivers. With a territorial area of 7,196 $km^2$ and a population of 1,154,654 people in 2011, South Moravia has a population density of 157 people per $km^2$, and is classified by Eurostat as 'significantly rural'. However, these figures include the capital city of Brno, which with a population of 371,371 is home to a third of the region's residents. Outside Brno, the population density falls to 110 people per $km^2$ and the largest towns are Znojmo (34,476 residents), Hodonín (25,240), Břeclav (24,052), Vyškov (21,720) and Blansko (21,108).[1] More than half of the rural population of South Moravia lives in small villages, such that the social geography of the region is markedly differentiated (Vaíshar et al. 2011). South Moravia is one of 14 administrative regions of the Czech Republic and contains seven districts (see Figure 11.1) and 673 municipalities, 21 of which have enhanced powers.

### *Economic and Social Conditions*

The South Moravian Region is an area with significant economic potential. The region contributes 10.5 per cent of the gross domestic product (GDP) of the Czech Republic as a whole, and GDP per head in the region was 76.7 per cent of the EU average in 2009. However, the regional economy is dominated by Brno, with its industrial tradition in manufacturing and product processing. The manufacturing sector contributes 21.2 per cent of gross value added (GVA) in the

1 Population figures are from *Oficiální stránky Regionálního informačního servisu*. Available at: http://www.risy.cz [accessed: 17 October 2011].

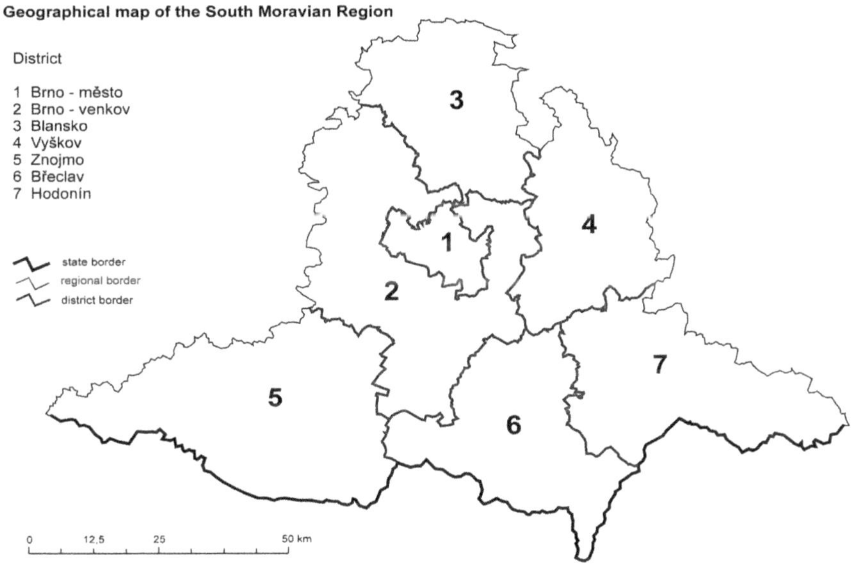


**Figure 11.1 Map of the South Moravian region**

region, compared with only 2.7 per cent contributed by agriculture (BusinessInfo. Cz 2011). Over the period of liberalization since 1990, however, both agriculture and manufacturing have decreased in importance relative to services. The share of regional employment in the primary sector, including farming, fell from 5.9 per cent in 1995 to 2.9 per cent in 2010, whilst that in the secondary sector decreased from 40.3 per cent to 35.7 per cent. Employment in the service sector increased from 53.7 per cent to 61.4 per cent of all regional employment (CZSO 2010). Relatedly, employment in large industrial enterprises has fallen more sharply in South Moravia than in other parts of the Czech Republic, but there were still 247 industrial enterprises recorded in 2010 with 100 or more employees (Portál Jihomoravský kraj 2011).

Unemployment increased with economic restructuring during the 1990s, but remained fairly stable during the early 2000s and had fallen back to just over 5 per cent in 2007(statistics from Eurostat). The global economic recession from 2008, however, has had a significant impact with total employment in the region falling by 9.4 per cent between 2009 and 2010 (Portál Jihomoravský kraj 2011). This impact has been uneven and there are notable geographical variations in unemployment rates, with the lowest rates in Brno and surrounding communes and the highest rates in rural communes in the south-west of the region. Some villages in the Znojmo Region had unemployment rates above 25 per cent in October 2009. Levels of economic inactivity are also higher in rural communes than in towns (Woods and Jones 2008).

These differences further reflect variations in both opportunities for employment and the capacity of the workforce. Many of the largest employers in the region, including hospitals, universities and schools, as well as Czech Railways, Czech Post and regional transport enterprises, tend to be located in Brno or in the larger towns, such that nearly three-quarters of the working population in rural municipalities commute outside their home town or village for work (Woods and Jones 2008). Equally, levels of educational attainment are higher in urban areas, with significant proportions of rural residents having only a basic education or vocational training without a secondary or higher qualification (ČSU 2009).

Between 2002 and 2006, €13 billion (349 billion CZK) was invested in South Moravia by foreign investors, of which 62 per cent went to Brno, 17 per cent to the peri-urban Brno-venkov district, and 21 per cent to the remaining rural districts (Czech National Bank, quoted by Woods and Jones 2011). During the fourth quarter of 2008, 6,085 business licences were awarded to non-Czech nationals in South Moravia. However, at 3.1 per cent of all business licences issued, this is below the national average for the Czech Republic, even if Prague is excluded. Yet, when international and domestic investments are combined, South Moravia has the highest rate of investment in the Czech Republic. In 2010, for example, 52 investors from outside the region (both national and international) made investments in South Moravia totalling 1.123 billion CZK and creating 1,678 jobs (CzechInvest Agency 2011).

Foreign investments have included the purchase of farmland and agricultural enterprises, and the arrival of international retail chains, especially in new retail parks. At the same time, major new retail parks such as Avion and Olympia have been located close to transport intersections and are designed to service a transnational market with custom from Slovakia. Other examples of foreign investment include the Czech Technological Park in Brno, established as a joint venture between Brno City and the British firm P&O, which has attracted transnational corporations including Siemens, IBM, FEI, SGI, Vodafone, Honeywell Controls and Symbol Technologies and the Central Trade Park at Modřice, developed by a Dutch company, which has also attracted international firms including Andrew Telecommunications, Danaher Motion, Tyco, IFE/Knorr Bremse and IMI Norgren (Woods and Jones 2008).

However, the international networks of smaller, rural-based enterprises in the region remain limited. Forty-five small and medium enterprises in six rural districts of South Moravia were surveyed as part of DERREG between 2009 and 2011, with 15 follow-up interviews. Less than 10 of these companies had significant international business relations, with most being focused on local or national markets and working primarily with regional or national partners. Where companies did have international links these were largely with businesses in neighbouring regions of Austria, Slovakia, Poland and Germany, and none had business transactions outside Europe. Obstacles identified by interviewees to developing international trade included organizational logistics, financial resources, transportation difficulties, language barriers (with only a few having knowledge of English or German) and the lack of visibility for and knowledge about Czech products in foreign markets.

*Policies for Regional Development*

Policies for the economic development of the region are articulated and implemented through parallel strategic and territorial development programmes. The Development Strategy for the South Moravian Region envisages a dynamically developing region which, in accordance with the principles of sustainable development, has a powerful and competitive economy based on modern technologies with high added value; high levels of education with an emphasis on preparation and use of quality human resources; maintains an acceptable state of the environment and landscape-oriented modern agriculture, maintaining regional cultural specifics; and uses its favourable geographic position on the major continental axis, using the potential of tourism and conceptually streamlines the process of suburbanization (Strategie rozvoje Jihomoravského kraje 2020, 2012).

The policy aims to stabilize the structure of medium and large businesses in the region, including simplifying legal restrictions on land ownership, but it also places a strong emphasis on integrated rural development, combining the modernization of traditional sectors with diversification and the valorization of cultural and environmental heritage with both economic and social goals. As such it seeks to increase the attractiveness of the countryside as a place to live through support for new environmental technologies, and the restoration of public buildings and spaces, landmarks and amenities (especially the reuse of existing building stock). At the same time it aims to generate economic growth by promoting rural energy independence (including biomass and production engineering of cereals for bioethanol), supporting handicrafts and traditional economic activities that use local resources and encouraging rural tourism and agri-tourism in areas of cultural and natural heritage, as well as investment by companies producing high technology products.

Accordingly, the strategy positions agriculture and forestry as peforming balanced production, environmental and social functions. Thus it involves investment in the modernization of farms, training, comprehensive landscaping and the development of ecological agriculture and forestry (as a supplement to intensive agriculture and forestry), as well as horizontal and vertical cooperation between businesses and other organizations, for example to support Moravian wine. Furthermore, the intention is not only to provide for regional and national markets, but also to develop niche exports to global markets of commodities including certified products of Moravian malt, pork and poultry, as well as wine. As such, the development of rural South Moravia in the context of globalization is closely tied to the valorization of its cultural and environmental resources.

### *Cultural and Environmental Resources*

Agricultural land covers nearly 60 per cent of the region's area, and 83 per cent of this is arable land, with agricultural production focused mainly on cereals, oilseed rape and sugar beet. There is a long tradition of local agricultural specialization, linked to specific local geographical and climatic conditions, which includes areas noted for wine, fruit and vegetable growing. More than 90 per cent of Czech vineyards are located in the region. In addition to arable farming, South Moravia is also a leading region for pig and poultry production in the Czech Republic (ČSU 2010).

These landscapes and practices form part of the culture economy of South Moravia that is increasingly being valorized as a resource for rural development, including diversification beyond agriculture. The tradition of viticulture exemplifies this, being deeply rooted in Moravia. Wine production in Moravia developed in the sixteenth and early seventeenth centuries, when vineyards covered more than 30,000 hectares – compared with a total of 18,710 hectares for the Czech Republic as a whole today. However, viticulture has been revitalized, both as a productive activity and as the basis for tourism. A series of Moravian Wine Trails have been set up as long-term projects under the auspices of the Partnership Foundation, with the aim of protecting cultural heritage and developing wine tourism. Each of the ten wine regions in South Moravia has its own range of wine trails, and all circuits are connected to the backbone of the Moravian Wine Trail leading from Znojmo to Uherské Hradiště. Over four years, 220 wine-growing villages have been involved in the projects and 1,090 km of cycling routes have been marked. Moravian wine trails are used by many municipalities and small businesses as a tool for sustainable tourism development and promotion of small business, with the economic benefit directly tied to respect for the landscape, preserving folk traditions and enhancing cultural heritage associated with the winery (VŠE Praha 2007). Hypothetically it is possible to consider a positive influence of folklore, culture and wine to the regional development of Moravian Slovakia (the cultural region in the South-East Moravia near Slovak and Austrian border).

More broadly, the importance of the natural environment of South Moravia has been recognized through international designations, which in turn have been catalysts for the development of tourism. The Podyjí National Park was established in 1991 covering 63km$^2$ of primeval forest in the Dyje river valley along the Austrian border. The corresponding Thayatal National Park was established at the Austrian side of the border in 2000. The Czech and Austrian authorities cooperate in the management and promotion of the area, signing an agreement on cross-border crossings by tourists hiking in the parks in 2006 and developing the Podyjí-Thayatal Visitor Centre through a partnership utilising INTERREG III funding (Woods and Jones 2008).

The South Moravian Region includes 203 Natura 2000 areas (8 Bird Areas and 195 major sites), which protect 9 per cent of its territory. There are additionally eight Natura 2000 Special Protection Areas designated in the South Moravian

Region and three Ramsar sites including the Lednice fishponds (designated 1990), Lower Dyje floodplain (1993) and the Punkva subterranean stream (2004). The Czech and South Moravian authorities have further designated three Protected Landscape Areas: Moravský kras (Moravian Karst), Pálava (Pavlovské vrchy Hills) and Bílé Karpaty (White Carpathian Mts.) and 282 small protected areas, including 175 sites of European value. The small protected areas include 18 national natural reserves and 13 national natural monuments. The total area of protected landscapes exceeds 50,000 hectares.[2]

The Pálava limestone hills in the south of the region were designated as a Protected Landscape Area in 1976 and as a UNESCO Biosphere Reserve in 1986. The reserve was extended in 2003 with the floodplain forests at the confluence of the Morava and Dyje rivers and the landscape of the Lednice-Valtice Cultural Landscape, forming the new Lower Moravia Biosphere Reserve (Dolní Morava), encompassing 24,200 hectares between Nové Mlýny reservoir and the Austrian and Slovakian borders. The area of the Biosphere Reserve includes a strictly protected 6,900 hectares of core zone, 9,000 hectares of buffer zone and 8,834 hectares of transition zone. Around 20,000 people live in the Biosphere area; mainly in the transition zone. The management plan includes objectives for both nature conservation and sustainable development, with a particular focus on sustainable tourism as in all UNESCO Biosphere Reserves. The Lower Moravian Biosphere Area also includes the Lednice-Valtice Cultural Landscape, which was separately recognised by UNESCO as a World Heritage Site in 1996.

Indeed, in addition to its environmental resources, South Moravia also boasts a notable cultural heritage. Amateur clubs, ensembles and groups operate in community houses of culture, music schools, and different elementary schools or as separate interests of civic associations. These provide the infrastructure for amateur cultural activities and are important not only for residents and civil society, but also for tourism and the presentation of the region abroad. There is particularly strong representation of folk ensembles in the region, maintaining traditions in folk music, songs and dances in areas around Břeclav, Strážnice, Hodonín, Podluží, Horňácko and Dolňácko. These are important markers of territorial identity, and the folk dance 'verbuňk' was officially entered in the UNESCO list of Masterpieces of the Oral and Intangible Heritage of Humanity in November 2005. Similarly, in the southeastern districts of the region, ethnographic festivals have served to preserve local traditions and folklore, whilst the region hosts a number of museums and galleries of various types.

The South Moravian Region is therefore an area with a rich and unique culture economy. It has considerable potential for developing its cultural resources for residents and visitors alike, and an important starting point for the fulfilment of this vision is to strengthen the linkage of the cultural offer of region to tourism (BusinessInfo.Cz 2011).

---

2 Information on protected areas is sourced from *Oficiální stránky Českého statistického úřadu*. Available at: http://www.brno.czso.cz/ [accessed: 17 October 2011].

## Globalization and Sustainable Development

As the section above has described, rural development strategies in the South Moravian Region are attempting to balance engagement with the global economy with preservation of the region's cultural and environmental resources on which its economic potential depends.

Discussion about future development of the South Moravian Region should therefore be directed towards sustainable development. This is a multidimensional process, which endeavours to integrate economic, socio-cultural and environmental goals in a sustainable manner (Moseley 2003). According to Griffiths et al. (2004), the transition to rural sustainability must be based on creating a good standard of living, the preservation and protection of fundamental life support processes and the use of renewable resources. The principal idea is to ensure balance among three basic pillars: social, economic and environmental. Whilst responsible behaviour and attitude towards the environment is essential for environmental sustainability (Šimková 2008), the economic pillar is focused on increasing competitiveness as well as on ensuring a sustainable growth of the regional economy (Stejskal and Kovárník 2009). This is more easily measurable than social sustainability as it can be defined in numbers (Munro 1995).

To achieve economic sustainability, rural development strategies in the South Moravian Region need to support the emergence of small companies and factories that can compete with big brands and transnational corporations that displace not only small and medium-sized enterprises but also established companies that respect local peculiarities and preferences (Lesáková 2001). Assistance should also be focused on the diversification of economic activities in rural regions (Hrabánková et al. 1996), including new uses for agricultural production, and in creating a suitable environment for residential and business investment that will create new job opportunities. These actions have been the objective of the Czech Rural Development Programme during the 2007–13 Structural Funds round (SZIF 2007).

Furthermore, the diversification of the rural economy can be managed to meet both economic and environmental sustainability goals. Thus, one of the most notable economic trends has been the gradual decline of agriculture caused by the unregulated overproduction of food, falls in sales of agricultural products, the dissolution of agricultural companies and the loss of an agricultural perspective within the regional population. Although this trend can be attributed in part to EU agricultural policy, it is also a common priority for EU countries to promote the sustainable management of food production, the use of renewable energy sources and the maintenance of rural cultural landscapes and villages. In the South Moravian Region this has included support for renewable sources of heat and electricity, including burning straw, hemp, willow and poplar wood, and biogas and biodiesel production from sources such as oilseed rape, along with the simultaneous treatment of organic waste, all providing a new focus for agricultural production (Sýkora 1998).

The broader challenge is to generate social capital in rural areas to provide opportunities for all residents to create their own healthy and viable countryside (Griffiths et al. 2004). This is related to the development of tourism in the South Moravian countryside, which is seen as being of considerable economic and social benefit to rural areas through the income and infrastructural developments that it may bring to marginal and less developed regions. Yet, in some cases tourism development may also bring impacts of congestion, intrusion and loss of amenity (Hall 2005). There is a need for strategic planning for the efficient use of resources and funds to balance economic, environmental and social sustainability, with involvement from stakeholders in policy formulation (Mitchell and Hall 2005).

In a global context, the goal of the development of a sustainable countryside is to attain sustainable economic growth and improve living conditions to ensure that rural regions will be attractive places for living, and will be able to make a positive contribution to the national economy (Woods 2011). In the condition of globalization, the sustainability of the countryside may also be understood as the maintaining of regional identity and cultural specificities as a counterbalance to the standardization of patterns of production and consumption. When related to the differential impact of globalization processes in the South Moravian Region, it is possible to say that due to globalization, core areas are developed but peripheral areas' development have been restricted and an already problematic situation has been deepened. This includes the increasingly complex consequences of migration, caused by the social and economic decline of marginal rural areas and globally uneven urbanization. In South Moravia this has been reflected in depopulation in peripheral districts and population growth in core areas, leading not only to social problems but also the disintegration of rural cultural (Suša 2003).

However, there is always a chance that marginal rural areas might be able to exploit their social and environmental capital. Peripheral rural areas are valued for their regional identity (traditions, culture and the environment), and distance from urban centres can be helpful for the survival of cultural diversity (Ferrão and Lopes 2004, Sedlacek et al. 2009). This potential might be exploited by rural peripheral areas creating regional conglomerations to support economic development (Epp 2002). The South Moravian countryside in the era of 'globalization' could contribute to present and promote local and regional traditions (Vaishar and Zapletalová 2010). Furthermore, preserving regional traditions and emphasizing the somewhat overlooked missing link between the global and the local (Moldan 2001) – that global processes and phenomena are composed of countless local processes and phenomena – would also help to address global environmental concerns, including the loss of biological diversity. In the next section, we highlight some businesses and initiatives in the South Moravian Region that are already working on these principles.

## Good Practice in South Moravian Responses to Globalization

It is clear that 'global problems' require 'global solutions' based on international cooperation by all countries, but a difference can also be made by successful local projects and interventions. In the South Moravian Region, a number of examples of good practice can be identified that are developing effective ways of responding to the challenges of globalization and promoting sustainable development (Šťastná et al. 2013).

### *Sustainable Forest Management*

Forests and woodland cover more than a quarter of the land area of the South Moravian Region, but the significance of forestry as an economic sector has declined. As around the world, forestry in the Czech Republic has come under pressure both from competition in the global forestry industry and from environmental concerns. These pressures have prompted a change in focus in forest management in the South Moravian Region, with a shift towards sustainable forest management practices, based on an ecosystems approach. This has involved thinking about how forestry can support both environmental and economic goals, both directly and indirectly.

One example of this is the renewal of windbreaks in South Moravia (Litschamnn et al. 2007). The main objective for planting windbreaks has been to reduce the risk of wind erosion and to retain moisture in the landscape. There was a 25 per cent increase in rainfall in the South Moravian Region in 2009 compared to the long-term normal precipitation 1961–90, which means that recent years have been relatively wet.[3] The 'Forests of the Czech Republic' State Enterprise has led the reconstruction of windbreaks so they could carry out all functions necessary for the local landscape.

Another good practice relates to the issue of sustainable forest management and education, aimed at providing forest visitors with targeted information about the purpose of sustainable forest management, the benefits and services provided by the forest to the human population and about the functioning of the forest ecosystem. This has been the purpose of a forest pedagogy programme introduced in 2007 and involves education on the positive relation to nature, showing human activities in the forest and understanding the importance of the forest for people. Forest pedagogy is the simplest and most natural way for the foresters to introduce the forest environment and tasks of forest management to the general public (Mezi stromy 2007).

3 Based on meteorological data from *Oficiální stránky Českého hydrometeorologického ústavu*. Available at: http://portal.chmi.cz/portal/dt?action=content&provider=JSPTabContainer&menu=JSPTabContainer/P4_Historicka_data/P4_1_Pocasi/P4_1_5_Uzemni_srazky&nc=1&portal_lang=cs#PP_Uzemni_srazky [accessed: 1 November 2011].

As a third example, forest products have been marketed in a way that emphasizes sustainability. In 1998, several NGOs in Brno (Rezekvítek, ZO ČSOP Veronica, Lipka, EkoCentrum Brno and YMCA) started selling Christmas trees in pots as an environmentally friendly form of Christmas décor (Hromádková and Vlašín 2004), with a further organization, 'Za sebevědomé Tišnovsko', joining this activity in 2007. People who want to contribute to nature conservation during Christmas can buy a living Christmas tree. In the spring, people have the opportunity to return it to the land in a common planting. In addition to reducing waste, the initiate also aims to return fir trees – once a widespread coniferous species but now depleted – to Czech forests, at least symbolically.

### *Znojmo Winery Local Action Group*

The Znojmo Winery Local Action Group (LAG) is a partnership of businesses, public sector agencies and NGOs aimed at supporting rural development in the Znojmo Region. It is a non-political community in which all partners have an equal vote, the large municipalities as well as small civic associations. It seeks to engage with local residents, believing that only the close cooperation and partnership of farmers and villagers are a guarantee of quality rural development.

The group brings income into local municipalities and businesses through its activities and also works with the non-profit sector to achieve social benefits. In particular, the LAG provides advice to its members and also non-members from the area, but mainly is responsible for bringing in financial resources to the district from the Czech Rural Development Plan. The LAG declares that this is in accordance with the basic strategic direction of the Regional Operation Programme of the NUTS 2 region 'Jihovýchod': Development of sustainable tourism and Sustainable development of towns and rural settlements.

The target beneficiaries are micro and small enterprises, focused on crop production and food processing, thus supporting the traditional agricultural sector. Entrepreneurs are motivated to participate primarily by the possibility of obtaining a subsidy or by getting assistance with product promotion. Some of them are contacted by the network; some of them approach the LAG first. The LAG can also help with grant applications, correcting errors and with reporting changes and the final account for grants received.

### *South Moravian Innovation Centre (JIC)*

The South Moravian Innovation Centre (JIC) helps innovative companies, students with original ideas, researchers and inventors promote collaboration between industry, research institutions and public administration. The centre creates a friendly environment for innovative business in the South Moravia Region. The main participants in the network activities are business owners who are in charge of growth value of the firms, investment and mergers, top managers of companies and acquisitions; chief strategy officers, who manage growth of strategic projects; marketing and sales managers and production managers.

The major JIC project is the 'innovative network academy'. The aim of the innovative academy is to train managers and business owners to face new challenges in a sharper competitive environment in which innovation is just becoming a new source of competitive advantage and society growth. The innovation academy program is based on a combination of best practice and trends in the field of innovation management, and also includes a practical element in the form of workshops and coaching of specific business projects. Interactive workshops address issues such as innovative business strategies, building powerful corporate innovation processes, new markets and products and increasing the efficiency of business processes. Participants also take part in excursions to leading companies in the region, and receive a certificate after the successful defence of an innovative project that they have developed at the end of the course.

JIC also provides other activities such as the incubation program, providing support for start-up companies with a business plan built on an innovation and offering potential return on investments. Companies selected for the incubation program get the best environment for starting a business. It includes assistance with creating a business plan and all company-based training, consulting in the areas of finance, marketing, management and human resources, networking and collaboration with universities in Brno, as well as access to accounting, tax and legal advisory services, incubators, scientific and technological parks, innovation vouchers, technology transfer and patent and licensing funds. Innovation vouchers are also offered as one-shot subsidies for entrepreneurs to develop pilot cooperation projects with a knowledge provider. The intention is to build up long-term relationships between both parties to be able to work together on challenging projects in the future. Innovation vouchers help on the one hand to increase the competitiveness of small and medium-sized enterprises and on the other hand to make more effective research commercialization at the research institutions.

Some of the JIC's activities are specifically aimed at assisting businesses in the South Moravian Region to engage with globalization by developing international networks. These brokerage activities involve meetings of international partners, including business representatives and academics. They usually focus on one or more disciplines and aim at establishing contacts and partners for international cooperation in research, development, production and technological applications. Participants have the opportunity to meet at pre-organized meetings, which are built on their interest in cooperating with other participants. For example, a two-day ICT Brokerage was held in Brno in April 2009, which was targeted at research institutes and companies from all over Europe, but especially from neighbouring regions. ICT Brokerage 2009 was attended by a total of 98 participants, including 35 foreigners. Most participants evaluated the event as being very beneficial; the highest rated being the possibility of finding a partner for a joint project and to make contacts relevant to their research project or activity.

## Conclusion

Globalization comes to the South Moravian Region through Brno, direct cross-border contacts and European Union policy. It brings both risks and opportunities for the region (see Jeníček 2003, Moseley 2003, Labrianidis 2004 and others). On one side, globalization has contributed towards the degradation of regional culture and identity, and the distancing of the regional economy and society's relationship to the landscape. On the other side, the South Moravian Region continues to maintain a local individuality through family traditions, living folklore and cohesion of local residents that form a base of cultural and environmental resources that could be developed to exploit the economic opportunities of globalization. This entails efforts to maintain solidarity and the preservation of the culture typical for this region.

The South Moravian rural landscape has been disturbed in the past, especially by intensive agriculture and its consequences. The economic sustainability of the South Moravian countryside is facing challenges from global economic transformation, competition in the world market and the implications of the CAP (Common Agricultural Policy). This means that there is a need to find activities that can ensure the prosperity of a market economy in the countryside, but which also contribute to social and environmental sustainability. As discussed in this chapter, regional development strategy in South Moravia has started to respond to this challenge by promoting forms of sustainable development that build on the region's distinctive cultural and environmental heritage, appealing to domestic and international customers alike, including the culture of wine and sustainable tourism.

As such, the South Moravian Region can be seen as an example of a region pursuing neo-endogenous rural development through the valorization of its 'culture economy'. As in many other rural areas in Europe, a particular focus has been developed on the valorization of nature through innovative projects and the rearticulation of knowledge resources (see also Siebert et al. 2008) to create products for both local and international markets.

Globalization is a reality and regions such as South Moravia need to find ways to benefit from it. The future will continue to be risky and uncertain as globalization is not yet settled as the fulfilment of a millenarian vision (Mezřický 2003). Despite these threats and risks, however, the opportunities presented by globalization also have the potential to promote sustainable development and support equitable and harmonious development in the South Moravian Region and elsewhere (Zjuganov 2002).

## References

BusinessInfo.Cz (2011) *Oficiální portál pro podnikání a export* [online]. Available at: http://www.businessinfo.cz/cz/clanek/jihomoravsky-kraj/charakteristika-jihomoravskeho-kraje/1000924/40555/#ekudaje [accessed: 9 October 2011].

ČSU (2009) *Postaveni venkova v Jihomoravskem kraji*. Available at: http://www.czso.cz/xb/edicniplan.nsf/t/41004D2F6B/$File/64136109.pdf [accessed: 17 October 2011].

ČSU (2010) *Statistická ročenka Jihomoravského kraje 2010*. Available at: http://miniscitani.cz/csu/2010edicniplan.nsf/t/F50030FC21/$File/64101110.pdf [accessed: 17 October 2011].

ČZSO (2010) *Český statistický úřad: Základní tendence demografického, sociálního a ekonomického vývoje Jihomoravského kraje 2010*. Available at: <http://www.brno.czso.cz/csu/2011edicniplan.nsf/krajp/641364–11-xb> [accessed: 31 October 2011].

Epps, R. (2002) Sustainable rural communities and rural development, in Bowler I.R et al. (eds) *The Sustainability of Rural Systems*, pp. 225–46. Netherlands: Kluwer Academic Publishers.

Ferrão, J. and Lopes, R. (2004) Understanding Peripheral Rural Areas as Contexts for Economic Development, in Labrianidis L. (ed.), *Future of Europe's Rural Peripheries*, pp. 31–61. Farnham: Ashgate.

Griffiths, J., Beckmann, A., Serafin, R., Veselý, M. and Kundrata, M. (2004) *Rural Livelihoods for Sustainability. Stories of rural regeneration from Central Europe*. Czech Environmental Partnership Foundation, Brno, Czech Republic.

Hall, D. (2005) Sustainable Rural Tourism and Rural Change, in Schmied, D. (ed.) *Winning and Losing. The changing geography of Europe's rural areas*. Farnham: Ashgate.

Hrabánková, M., Trnková, V. and Kleinertová, D. (1996) *Hodnocení území z pozice agrární regionální politiky a rozvoje venkova*. Výzkumný ústav zemědělské ekonomiky. Praha.

Hromádková, K. and Vlašín, M. (2004) *Vánoční stromky*. Brno: Veronica.

Lehmannová, Z., Jeníček, V., Chlapcová, I., Suša, O., Eichler, J., Kortusová, P. and Stuchlík, J. (2003) *Aktuální otázky globalizace*. 1. vyd. Praha: Nakladatelství Oeconomica, 407.

Labrianidis, L. (ed.) (2004) *Future of Europe's Rural Peripheries*. Farnham: Ashgate.

Lesáková, D. (2001) Vplyv globalizácie na tvorbu marketíngových strategií, *Marketing a komunikace 2/2001*, 6–8.

Litschamnn, T., Rožnovský, J. and Podhrázká, J. (2007) Využití optické porostity ke klasifikaci větrolamů, in Střelcová, K., Škvarenina, J. and Blaženec, M. (eds) *Bioclomatology and Natural Hazards*. Available at: http://cbks.cz/SbornikPolana07/pdf/Litschamnn_at_al.pdf [accessed: September 2011].

Mezi stromy (2007) *Lesnicko-dřevařský vzdělávací portál*. Available at: http://www.mezistromy.cz/cz/vzdelani/lesni-pedagogika [accessed: September 2011].

Mezřický, V. (2003) Peripetie procesu globalizace, in Mezřický V. a kol., *Globalizace*, pp. 9–29. Praha: Portál.

Mitchell, M. and Hall, D. (2005) Rural Tourism as Sustainable Business: Key Themes and Issues, in Hall, D., Kirkpatrick, I. and Mitchell, M. (eds) *Rural Tourism and Sustainable Business*. Great Britain: Cromwell Press.

Moldan, B. (2001) *(Ne)udržitelný rozvoj, ekologie, hrozba i naděje*. Praha: Karolinum.

Moseley, M.J. (2003) *Rural Development. Principles and Practice*. London: Sage.

Munro, D.A. (1995) Sustainability: Rhetoric or Reality? in Tryzna T.C. (ed.) *A Sustainable World. Defining and measuring sustainable development*, pp. 27–35. USA: International Center for the Environment and Public Policy.

Portál Jihomoravský kraj (2011) *Portál Jihomoravský kraj*. Available at: http://www.kr-jihomoravsky.cz/Default.aspx?ID=10529&TypeID=1 [accessed: 9 October 2011].

Ray, C. (1998) Culture, intellectual property and territorial rural development, *Sociologia Ruralis*, 38, 3–19.

Ray, C. (1999) Endogenous development in the era of reflexive modernity, *Journal of Rural Studies*, 15, 257–267.

Ray, C. (2006) Neo-endogenous rural development in the EU, in Cloke, P., Marsden, T. and Mooney, P. (eds) *The Handbook of Rural Studies*, pp. 278–91. London: Sage.

Sedlacek, S., Kurka, B. and Maier, G. (2009) Regional identity: a key to overcome structural weaknesses in peripheral rural regions? *European Countryside*, 1 (4), 180–201.

Siebert, R., Laschewski, L. and Dosch, A. (2008) Knowledge dynamics in valorising local nature, *Sociologia Ruralis*, 48, 223–39.

Šimková, E. (2008) Udržitelný rozvoj venkova a venkovská turistika. *Ekonomie E+M*, Nr. 1/2008, 26–32.

Stejskal, J. and Kovárník, J. (2009) *Regionální politika a její nástroje*. Praha: Portál.

Suša, O. (2003) *Sociální globalizace a její sociální aspekty*. Praha : *Oeconomica*, 2003.

Sýkora, L. (1998) *Venkovský prostor 2. díl: Územní plánování vesnice a krajiny*. CVUT, Praha, 156 s.

SZIF (2007) *Program rozvoje venkova Česke republiky na obdobi 2007–2013*. Available at: http://www.szif.cz/irj/portal/anonymous/CmDocument?rid=%2Fapa_anon%2Fcs%2Fdokumenty_ke_stazeni%2Feafrd%2F1180428724933.pdf [accessed: 9 October 2011].

Šťastná, M., Toman, F., Vaishar, A., Woods, M. and Vavrouchová, H. (2013) European Rural Regions in the Era of Globalization: The South Moravia Case *Innovation: The European Journal of Social Science Research*. 26. 4, 354–64.

Vaishar, A., Šťastná, M., Vavrouchová, H., Stejskal, B., Hlisnikovský, L., Jakešová, L., Kniezková, T., Lipovská, Z., Ševelová, M., Novotná, K., Ošťádalová, Z., Kozáková, H. and Marciánová, G. (2011) *Současný stav a vývojové tendence jihomoravského venkova*. Brno: Mendelova univerzita v Brně.

Vaishar A. and Zapletalová, J. (2010) Sustainable development of the Moravan countryside. *Revija za geografijo*, 5–1, 43–54.

Van der Ploeg, J. D. and Marsden, T. (eds) (2008) *Unfolding Webs: The dynamics of regional rural development*. Assen: Van Gorcum.

VŠE Praha (2007) *Venkovský cestovní ruch, jeho specifika a podmínky pro rozvoj*. Available at: http://www.mmr.cz/CMSPages/GetFile.aspx?guid=68c498e0–3dbb-4f2d-9c8b-c336435253b7 [accessed: 9 October 2011].

Woods, M. (2011) *Rural*. London and New York: Routledge.

Woods, M. and Jones, L. (2008) *DERREG Case Study Contextual Report 1: Jihomoravský kraj*. Available at: http://www.derreg.eu/system/files/Jihomoravsky%20Kraj%20WP5%20contextual%20report.pdf [accessed: September 211].Zjuganov G.A. (2002) *Globalizace slepá ulička nebo východisko?* [Říčany] OREGO, 36 s. ISBN 80-86117-82-0.

# Chapter 12
# Globalization Processes and the Restructuring of Europe's Rural Regions

Michael Woods, Birte Nienaber and John McDonagh

## Introduction

The chapters in this volume have illustrated how the challenges and opportunities presented by globalization have been encountered and experienced in selected rural regions in different parts of Europe. The evidence presented has demonstrated that contrary to some popular perceptions, rural areas are just as entangled and embedded in global networks as cities, forming an emergent yet uneven 'global countryside' (Woods 2007). The multidimensional character of globalization (Aggarwal 2006, Kelly 1999, McCarthy 2008, Steger 2003) and its mediation through regional contexts and capacities – as described in Chapter 1 – means that no two regions experience globalization in the same way. Nevertheless, there are over-arching processes and common impacts that can be observed in multiple regions, allowing some general conclusions to be drawn. Furthermore, there is potential for rural regions to learn from each other in developing responses to the challenges and opportunities of globalization, and for these lessons to be incorporated into the policies and programmes of regional authorities, national governments and the European Union.

In this concluding chapter, we pull together research findings from case study regions in the DERREG project, as presented in the preceding chapters. From these we highlight some key conclusions for each of the themes examined in DERREG – the embeddedness of rural businesses in international networks; international migration to rural regions; sustainable development and environmental capital in rural regions; and regional learning and capacity building – including observations on dynamics shaping the differentiated geography of the global countryside. Picking up this last aspect, we move on to postulate a typology of the positioning of rural regions in relation to globalization and identify the opportunities and challenges associated with these categories. Finally, we consider responses to globalization and propose recommendations for policy development.

## Globalization Processes and Rural Regions

As noted above, globalization is a multidimensional process and as such the processes impacting on rural regions are multiple and diverse. Although both academic studies of globalization in a rural context and policy statements on the same have tended to focus on economic phenomena – notably trade liberalization, the stretching of global commodity chains, the expansion and integration of transnational corporations and the circulation of international finance capital – the chapters in this volume have highlighted the equal importance of social, cultural and ecological dimensions of globalization for rural regions. These include the intensification of transnational mobility and migration, the homogenization and hybridization of cultural and consumption practices and the growth in consciousness of global environmental issues.

Moreover, the differential experience of globalization is further nuanced by questions of scale. If we follow Steger's (2003) description of globalization as the creation, multiplication, stretching and intensification of interdependencies and exchanges, it becomes evident that globalization is defined by a direction of travel rather than a destination. Relations do not need to be fully global in scope to be considered as part of globalization. For example, several of the regions discussed in this volume are located on international borders that have changed in character and meaning over recent decades. The expansion and intensification of economic networks and migration flows across these borders is hence an expression of globalization as much as ties and connections to places on different continents. Equally, the effects of globalization that are experienced in individual rural regions might be indirect as well as direct. For instance, pressure on farming in many parts of rural Europe has come in the main not directly from competition from imports or international 'land-grabbing', but from reforms of the Common Agricultural Policy initiated as part of EU commitments in World Trade Organization negotiations. Similarly, global pressures on environmental practice may be encountered primarily in the shape of national policies or campaigns that are responding to global issues. In other regions, it can be the rural development responses to globalization, including efforts to reconstruct local networks and systems that are the most evident yet indirect consequences of globalization.

With these qualifications in mind, the sections below briefly summarise the key findings from across the case study regions discussed in this book for the three themes of the embeddedness of rural businesses in international networks; international migration to rural regions; and sustainable development and environmental capital in rural regions.

### *Global Engagement and Local Embeddedness of Rural Businesses*

The engagement of businesses in an increasingly globalized economy is one of the most prominent aspects of globalization, and is often presumed to be disadvantageous to rural areas due to the smaller scale of rural businesses and the more limited scope for clustering of related industries. However, business networking can act as a surrogate for agglomeration, allowing dispersed rural firms to compete with those operating in an urban or suburban environment. The preconditions for success of rural business networks relate not only to physical infrastructure or availability of good access to information technology (such as high speed broadband), but also to a range of 'intangible assets'. These include human capital (education, age structure of the workforce and so on), aspects of social capital (an outward looking, non-risk-averse entrepreneurial culture, trust, cooperative ethos), facilitative governance arrangements and so on. It is also commonly argued that the most effective and dynamic business networks are those which combine dense and highly interactive patterns of local interaction with a smaller 'external' component, as captured in Granovetter's (1985) concept of the 'strength of weak ties' and Bathelt et al.'s (2004) emphasis on 'local buzz and global pipelines' (see also Copus et al. 2011).

The significance of such networks can be observed in evidence from the five regions in which this aspect of globalization was explored through the DERREG research: Swedish Norrland, Westerkwartier, Alytus county, Goriška and South Moravia (see Chapters 2, 3, 7, 8 and 11). A degree of internationalization of business transactions was observed in all five regions, but with significant variation. The highest levels of international engagement were recorded in Slovenia and Sweden, and the lowest in Lithuania and the Netherlands, with survey respondents in the Czech Republic exhibiting a bipolar distribution between internationally active firms and enterprises trading on a largely regional scale (see also Dubois et al. 2012). These different patterns reflect a number of factors including sectoral structure, the size of the regional market, human capital (especially language skills), local planning policy and proximity to national borders. Nevertheless, in spite of the observed differences, the underlying trend is clearly one of SME networks in rural regions opening up to interact beyond regional and national borders. Many respondents regarded internationalization as an important step in their business development, with improved transport and communications infrastructure helping to facilitate this expansion.

In addition to sales transactions, interactions with other companies are very important in product and process development, as well as for market consolidation and expansion. These interactions are especially important for SMEs that are too small to afford to fully internalize activities such as research and development or marketing. In contrast to transactional linkages, the survey data from the five regions shows that non-market interactions continue to be largely concentrated within the home region or country, most commonly with other SMEs. Transnational non-market linkages are more common in the Slovenian and Swedish case studies, but even here significantly less than the density of international transactional networks. As such, the results question the assertion in the existing literature that transactional and non-market linkages are closely aligned (see for example Tödling and Kaufman 1999), and point instead to the importance of 'institutional proximity' in building trust, and the role of face-to-face communication.

Furthermore, international linkages observed were predominantly with SMEs in other regions within Europe. This finding contrasts with the presumption in the existing literature that small firms are primarily enrolled in international networks through 'vertical integration' into global value chains controlled by large transnational corporations, usually as a supplier. Our evidence points instead to rural businesses developing direct networking with SMEs that occupy similar positions in the production chain in other countries, thus advancing a model of 'translocal integration' rather than distinctively 'global integration'. This may reflect the nature of products traded by rural enterprises, and the success of rural firms in carving out business spaces outside the sphere of large transnational corporations.

However, there are also notable variations in the patterns of international engagement by rural businesses in the five regions. Firstly, cross-border networking with adjacent regions in neighbouring countries contributed to the degree of internationalization in Slovenia (see Chapter 8, also Potočnik Slavič 2011), and to a lesser extent Sweden (Chapter 2, also Copus et al. 2011). In contrast, the border locations of the Lithuanian and Czech case study regions appeared to present fewer opportunities for cross-border interaction (see Chapters 7 and 11, also Šťastná et al. 2011). Secondly, proximity to domestic markets is a negative factor in stimulating internationalization – firms in regions with relative ease of access to sizeable domestic markets, as in the Dutch case study, have less of an incentive to develop international linkages than firms in more remote regions with smaller domestic markets, as in the Swedish case study.

Thirdly, variations in national and regional policies can have an effect. Spatial planning policies in the Netherlands, for example, militated against the development of internationally engaged businesses in the rural Westerkwartier, whilst policies in the Czech Republic are more favourable to industrial investment, possibly contributing to the higher degree of enrolment in vertically integrated networks observed for firms in the region. Similarly, the emphasis of business support through network brokering activities varied, with network brokers in Sweden and the Netherlands focusing on connecting firms to intra-regional and international networks, and network brokers in Lithuania, Slovenia and the Czech Republic emphasizing regional capacity building and access to public funding.

### *International Mobility and Migration to Rural Regions*

Increasing international migration is a feature of several of the regions discussed in this book, but with differences in the scale, direction and type of migration concerned. Examples including amenity or lifestyle migration to Pomurje and Swedish Norrland; seasonal or long-term labour migration to County Roscommon, Oberlausitz, Swedish Norrland, Saarland and Wales; return migration to County Roscommon and, to a lesser extent, Alytus county; refugee settlement in Swedish Norrland and the Westerkwartier; cross-border commuting in Saarland; and repeat emigration from Alytus county. These dynamics of international migration, and the narratives of migrants, were examined in detail in six of the regions: Swedish Norrland, County Roscommon, Alytus county, Pomurje, Oberlausitz and Saarland (see Chapters 2, 4, 7, 8, 9 and 10).

A multidimensional analysis of migration dynamics in these regions (following Düvell 2006) shows that the majority of foreign homeowners, migrant workers and return migrants moved from urban areas to the rural case study regions. Foreign homeowners and return migrants were predominantly from other EU countries, but migrant workers included significant numbers of non-EU citizens, including a majority of those interviewed for the study. Most migrants interviewed, of all types, had been or were intending to be long-term residents in their destination region (staying for at least 12 months). This included British property owners in the Pomurje Region of Slovenia, but not generally foreign homeowners in northern Sweden, where properties tended be owned as holiday homes and occupied for only part of the year.

Most of the migrants interviewed for the research were relatively long-distance migrants, including migrant workers from India, Sri Lanka, Vietnam, Mozambique and Central Asia in Germany, and return migrants from Canada and the USA in Ireland and Lithuania. Significant trans-European patterns of mobility were also evident, including foreign homeowners from Britain in Slovenia and from Germany in Sweden, and return migration from the UK and the Netherlands to Ireland and from the UK and Ireland to Lithuania. The major exception was Saarland, where the predominant pattern is of short-distance, cross-border migration from Luxembourg into border communities, driven by differential property prices (Chapter 10). However, whilst the geographical distance of migration was often substantial, migrants frequently stressed the importance of cultural proximity in their selection of migration destination.

The choice of destination was also influenced by political dimensions. Different labour market policies, and linked immigration controls, were the most commonly cited reason for the selection of destination region by migrant workers; however other factors including residential market, social and public regulations, environmental conditions and cultural ties between countries were also noted. For foreign homeowners, residential property markets were the most important pull factor, combined with environment and cultural ties. Interestingly, research in Slovenia also revealed the role of agents in promoting and facilitating international property purchases, thus consolidating bilateral migration relationships, in this case between Britain and the Pomurje Region (Chapter 8, also Lampič and Mrak 2012).

The majority of migrants interviewed in the research were integrated into the everyday lives of their communities, and reported positive experiences of settling in their destination regions. Although some described initial difficulties with communication and adapting to new cultural norms, as well as frustration with unfamiliar bureaucracies, most had established social networks and many were active in local community organizations (especially foreign homeowners and return migrants). In some communities, particularly in Slovenia, international migrants were welcomed by local populations for renovating empty properties and contributing to communities that had been in decline (Lampič and Mrak 2012). The perception by established local residents of international migrants is hence important in shaping experiences of integration. The most difficult problems were reported by international migrants in eastern Germany, especially those in former industrial towns with high levels of unemployment, who had encountered discrimination and marginalization (Chapter 9, also Nadler 2012, Nadler et al. 2012).

Difficulties with integration can also follow from the expectations of migrants. Some return migrants in Ireland, for example, discussed finding that they had little in common with old friends or feeling that they had been let down by a lack of family support, though most emphasized the importance of becoming involved in local organizations as a counter to the 'outsider' status of the returnee (Farrell et al. 2012), reinforcing earlier observations by Ni Laoire (2007) (see Chapter 4). A different dynamic of elective non-integration was identified for Luxembourgese migrants to Saarland, with a tendency to form social and geographical Luxembourgese enclaves reflecting the motivation of many cross-border migrants of escaping the perceived multicultural environment of Luxembourg (Chapter 10, also Frys and Nienaber 2011; Nienaber and Kriszan 2013).

Patterns and experiences of economic integration also vary. The majority of migrants interviewed across the regions were in employment, and most were satisfied with their employment situation. However, the types of jobs held varied by region and migrant group. Foreign migrant workers in Saarland tended to hold lower-grade jobs, and several reported problems finding suitable employment (Chapter 10, also Nienaber and Frys 2011). In contrast, a number of the migrants interviewed in eastern Germany were professionals who regarded their stay in the region to be a temporary staging post in their career progression (Chapter 9). Many migrants to rural regions have professional qualifications or skilled trades, but the utilisation of these skills varies. Return migration to Ireland has been facilitated in part by professional opportunities created by economic growth, including foreign direct investment, but entrepreneurship is also common (see Farrell et al. 2012). Around half of the return migrants interviewed in Ireland had started their own business on return, often using skills and experiences acquired abroad, yet return migrants to Alytus county, in Lithuania, described problems with translating skills and competences gained abroad into employment back home and difficulties finding appropriate jobs in the region (see Chapter 7).

Self-employment is also common among foreign homeowners or amenity migrants and may involve working or networking internationally, but can also represent a contribution to the local economy. Support for new business development is however variable. A scheme run by the municipality of Jokkmokk in Sweden is expressly aimed at encouraging entrepreneurship among international migrants, but return migrants in Ireland described mixed experiences of accessing support. The greatest challenges were faced by return migrants to Alytus county in Lithuania, where the economic situation presents few opportunities. Much return migration to Alytus was to some degree involuntary, usually for health or family reasons, and many return migrants interviewed struggled to find appropriate employment. Accordingly, return migrants to Lithuania frequently fail to stay, creating a pattern of repeat emigration (see Chapter 7).

*Environmental Capital and Sustainable Rural Development*

Programmes for sustainable rural development arguably represent an example of the indirect impact of globalization in rural areas. Global environmental issues such as climate change are instrumental in shaping the parameters of regional development, positing perceived new opportunities for rural regions in fields such as renewable energy production, but also presenting challenges for regions with a traditional economic base in activities such as mining or intensive agriculture. At the same time, regional development strategies need also to respond to local environmental concerns, such as pollution or landscape despoilment, which can either amplify or conflict with global concerns. These dynamics were explored through research in four of the regions discussed in this book: County Roscommon, Oberlausitz, Saarland and South Moravia (see Chapters 4, 9, 10 and 11, also Kriszan and Burdack 2013).

Sustainability has been adopted as a central theme in the regional development strategies of all the regions, reflecting the expectations of European and national policies, and the near-consensus that emerged from interviews with public, private and third sector stakeholders that sustainability and economic development can be mutually accommodated, and that protecting the environment presents opportunities for the regions concerned. In framing these policies there is often acknowledgment of the context of global challenges such as climate change, but such references tend to be limited and general. In terms of developing actual policies and initiatives, local environment concerns appear to be more important drivers. There is also a trend of movement from more general statements and prescriptions concerning renewable energy and organic agriculture in the early 2000s towards more focused policies and initiatives in later documents.

Thus, the global discourse of sustainable development is being translated into regional policies as mediated through an assessment of local problems, opportunities and pressures. Critically, the process of mediation places regional actors at the forefront of shaping regional development strategies. When asked to name the most influential actors in shaping regional strategies for sustainable development, relatively few of the stakeholders interviewed identified international bodies, national government departments or transnational environmental NGOs. Instead, stakeholders in all the case study regions pointed to local authorities and politicians, national and local conservation groups, local business and the regional media, as well as to local campaign groups, including both environmental movements and protest groups hostile to elements of sustainable development or environmental protection. As such, in mediating the translation of sustainable development discourse into policy, regional actors are not only localizing strategies, but may also be contesting elements of the global discourse.

Thus, conflicts can develop around the implementation of sustainability initiatives, especially where these are perceived to threaten established local interests. For instance, the Bliesgau UNESCO Biosphere reserve in Saarland is typical of an environmental designation that is modelled on international programmes or standards, framed by reference to global environmental concerns and awareness, but also promoted on the basis of potential opportunities for regional sustainable development – in this case the branding of sustainable agricultural products. Yet, in introducing restrictions on land management practices and hunting, the biosphere provoked opposition from some local residents, who perceived it as outside interference in established local practices (see Frys and Nienaber 2011). Similarly in the Görlitz district of Direktionsbezirk Dresden, regional development objectives of phasing out the heavily polluting historic brown coal industry and of forging a trans-national energy region based on renewable energy production, especially wind power, have been challenged by local support for mining and coal power stations as a source of employment and opposition to wind farms. This has led to a diluting of policy statements on wind energy and the ongoing operation of the last brown coal pits and power station, in spite of the direct involvement of Greenpeace in anti-coal protests (see Chapter 9).

At the same time, exploitation of a region's environmental capital can be a key part of regional development responses to globalization, contributing towards the development of an 'eco-economy' through the deepening, broadening and regrounding of economic activities based on environmental resources (Kitchen and Marsden 2009). In some cases these projects can use endogenous environmental capital to attract international capital – for example, Facebook's decision to construct a server plant in Swedish Norrland was influenced in part by the abundance of cheap, clean hydro-electricity. In others, environmental capital can be valorized as part of the reassertion of local networks to counter the centralizing tendencies of globalization. The *Bliesgau Genuß e.V.* (Bliesgau Consumption Association) in Saarland and the *Diši po Prekmurju* (Scent of Prekmurje) project in Slovenia both represent examples of locally focused 'deepening' activities, adding value through the promotion of sustainably produced regional food.

## A Typology of Regional Responses

The findings outlined above demonstrate the complexity of globalization impacts in rural regions, with the various case studies revealing examples of both positive and negative effects (see Table 12.1). Moreover, the analysis emphasizes the uneven geography of globalization in rural regions, with the same global processes leading to negative and positive outcomes in different regions, and variations between regions in whether the net impact is predominantly positive or predominantly negative. As discussed in conceptual terms in Chapter 1, variations between regions can be explained by their regional context and capacities, the specific catalysts operating in/on the region and the responses of local, regional

**Table 12.1 Positive and negative impacts of globalization as identified in the study regions**

| Positive impacts | Negative impacts |
|---|---|
| Access to international markets | Increased competition in domestic markets |
| Opportunities for business collaboration | Dependence on external market conditions |
| Investment of financial capital | Pressure to reduce costs |
| Expanded markets for endogenous products | Loss of local control of businesses |
| Multiculturalism from migration | Pressure on services from migration |
| Arrested demographic decline | Inflation of property prices |
| Labour supply in shortage areas | Social and cultural tensions with migrants |
| Support for sustainable development | Pressure on established rural industries and practices from environmental campaigns |
| Added protection for environments | Dilution of regional identity |
| Increased international tourism | Increased peripherialization in some regions |
| New cultural influences | |
| Opening of borders and reduced peripherality | |
| Coherence through social engagement | |

and national authorities and grassroots actors. Although the precise configuration of these dynamics will be different in each region, by analysing the evidence collected through the DERREG research and presented in this volume we are able to posit a broad typology of rural regional responses to globalization as outlined below (see also Woods 2014). We indicate where these types map onto the case study regions discussed in this volume, but we have also extrapolated from the case study research and drawn on statistical analysis used to contextualize the DERREG case studies to produce a more generalized typology that is also transferable to other regions elsewhere in the world. Individual regions might furthermore exhibit elements of more than one of these responses.

*Global resource providers*: regions with rich mineral and energy resources have benefited from the global resource boom, and particularly demand from East Asia, with local industries becoming strongly integrated into global networks. This transformation is usually driven by external corporate actors, with wealth flowing into the region, but at the cost of regional economies becoming vulnerable to political and economic trends, events and decision-making in distant countries. The challenge for regional development policy in this context is to maximize the retention of wealth generated by resource-exploitation within the region, and use this to enhance regional resilience by developing more autonomous, locally-embedded economic activities.

Among the regions examined in this volume, the 'global resource providers' model is evident only in Swedish Norrland, where significant mineral reserves are being mined to supply global markets. The mining industry in Swedish Norrland has benefited from the global resources boom, but with the consequence of increasing disparities in income between municipalities within the region. State-ownership of the major mine operator and the presence of a steel mill assist in retaining some added value in the region, and connections to China founded in part on the global resource economy are being consolidated to support wider regional development.

*Global farmlands*: regions with strong agricultural industries that are not only sufficiently resilient to adapt to competition from agri-food imports and the withdrawal of production subsidies, but have also become integrated into global agri-food commodity chains exporting produce to other parts of Europe and beyond. Agriculture in these regions is typically productivist and industrialized, and in addition to engagement in global commodity chains may also exhibit globalization impacts in the presence of transnational agribusiness corporations and reliance on foreign migrant labour. In some cases, 'global farmlands' have attracted foreign direct investment in the form of industrial farm units (such as intensive livestock farming), possibly by offering relaxed environmental regulations. In common with 'global resource providers', 'global farmlands' are potentially vulnerable to global economic trends and distant political events and corporate decision-making, and also present challenges in terms of realizing wider regional economic benefits from the success of one industry. None of the regions studied in this volume strongly exhibited a function as 'global farmland', although some included examples of export-oriented agriculture, especially viticulture.

*Branch plant economies*: attracting foreign direct investment in branch plants has been a common strategy for rural regional development, and can boost regional economies by generating new income, creating jobs, upgrading labour markets and supporting secondary businesses as suppliers and service providers. Regional development policy can make a significant difference in attracting foreign direct investment by consolidating and enhancing locational attractions by offering financial incentives (such as targeted tax-breaks), and appropriate infrastructure (such as pre-built business parks). However, as with global resource providers, branch plant economies are highly integrated into global networks and vulnerable to distant economic events and corporate decision-making. As such, regional development policies may need to address the retention or replacement of branch plants, or to support the parallel development of endogenous businesses.

The doubled-edged consequences of a 'branch plant economy' approach are demonstrated in County Roscommon. Attracting foreign direct investment had been a central element of Irish regional development strategy during the 1980s and 1990s, with County Roscommon benefiting from investment in a number of branch plants of transnational corporations, particularly in the medical instruments sector. However, the more recent economic downturn has resulted in job losses at some of these plants. Branch plant investment is also evident in case study regions in the Czech Republic, Lithuania and Slovenia, as is the downscaling of traditional industries such as textiles under pressure from global competition.

*Global playgrounds*: amenity-rich regions, especially those set in spectacular natural landscapes, have the potential to attract significant numbers of international tourists and amenity migrants to boost the local economy. Whilst the catalysts for the 'discovery' of new regions by international tourists are frequently external (including tour operators, airlines and the travel media), regional development policies can be significant in supporting international promotion and providing appropriate infrastructure. Challenges also exist for regional development in ensuring that tourism-based economies are sufficiently resilient to withstand seasonal variations and fluctuations in travel fashions, as well as in pre-empting and resolving potential tensions over the integration of amenity migrants, increasing property prices and pressures on affordable housing and environmental impacts. Whilst not on the scale of better-known resort areas in the Alps or on the Mediterranean coast, the attraction of British amenity migrants to Pomurje in Slovenia and of German holiday-homeowners to northern Sweden represented limited expressions of 'global playgrounds'.

*Niche innovators*: niche innovation is a response to globalization driven by internal regional actors and catalysts, and involves exporting specialist products based on endogenous resources to niche markets internationally; or attracting international visitors to niche events and attractions. Examples include the export of wild berries, Scandinavian bread and vernacular craft products from Swedish Norrland and tourist attractions in Swedish Norrland such as the 'ice hotel'. Regional development policies can help to stimulate and support niche innovation by identifying potential markets and working with entrepreneurs to develop business networks and train specialist employees.

*Trans-border networkers*: for border regions, the most significant globalization impacts may be associated with increased cross-border flows. These may include new business networks and trading opportunities for regional enterprises, but also investment in new factories, offices and retail facilities to service emergent transnational regions, and inter-regional migration associated with cross-border commuting. Regional development policies need not only utilize and consolidate opportunities for cross-border cooperation, but also need to protect regional interests within emerging transnational configurations – responding to risks that districts could lose businesses to centres with more favourable local policies or infrastructure across the border, or could become dormitory settlements for cross-border employment centres.

*Global conservators*: in regions with outstanding natural environments, some of the most significant globalization impacts can be the designation of protected areas following international models and classifications, such as UNESCO Biosphere Reserves, national parks and Natura 2000 sites. Many of these designations recognize social and cultural assets as well as environmental assets, and regional development agents increasingly recognize the opportunities arising from such international 'branding' for sustainable development, including sustainable tourism. At the same time, such designations can place restrictions on land use and management, farming and hunting, potentially requiring a reorientation of local economic activities. The 'global conservator' approach has been observed on a geographically limited scale within some of the regions in this volume, as well as protected areas in Slovenia and the Dresden Region of eastern Germany, with efforts to link environmental protection with sustainable development initiatives (Chapter 9).

*Re-localizers*: whilst the above responses have focused on engaging international networks and opportunities, an alternative response to globalization in rural regions has been to reassert local networks and cultures. In extreme cases this can include resistance to globalization impacts, such as opposition to new factories or tourist resorts or the sale of local companies or strategic properties or hostility towards immigrant groups. More commonly, however, it involves the reconstruction and strengthening of local value chains, for example through initiatives such as farmers' markets to consolidate local food networks, the rediscovery of local craft products or alternative regional 'currencies' aimed at retaining money within the locality. Regional development policies and grassroots initiatives both play an important role in initiating and supporting such projects.

Examples of a 're-localizing' strategy were observed in the Comarca de Verín in Spain, including initiatives to revitalize the historic chestnut economy; protect rare regional cattle breeds by marketing their beef as a premium product; and resurrect traditional carpentry craft production. However, these examples also illustrate the entanglement of even re-localizing strategies within wider global networks – the carpentry initiative, for example, whilst producing furniture historically distinctive to the region has had to import wood from France, Russia, Scandinavia and North America because the local *monte* forest has become too degraded to supply its requirements (Chapter 6). Aspects of re-localization are also evident in the projects developed through regional learning in the Westerkwartier (Chapter 3), as well as in the parallel study of Wales discussed in Chapter 5.

The variations outlined above form a typology of responses rather than a typology of regions because frequently more than one response can be found in a single region. Indeed, the different responses involve different balances between short-term economic effects and long-term sustainability, as reflected in the degree of integration into global networks and therefore vulnerability to external events. Figure 12.1 plots these relationships along the axes of 'economic impact' on a region to a response strategy – extending from a 'positive' boost to investment, GDP and employment to a 'negative' impact of limited growth or loss of jobs and income – and the 'vulnerability' of a region's economy to external events and decision-making. Thus, resource-provider regions have a strong positive impact in terms of GDP and job creation, but are highly vulnerable to fluctuations in demand and price for commodities in distant markets. Trans-border networking can also have a strong positive impact from business relocations and expanded markets, but is less vulnerable to external events because its international connections are over a shorter distance and may be across multiple sectors. A re-localizing strategy produces more modest economic returns in terms of investment and job creation, but by increasing local economic control is more resilient to external pressure. Finally, branch plant economies have been plotted with a sliding position because the initial investment has a strong positive impact for a region, but they

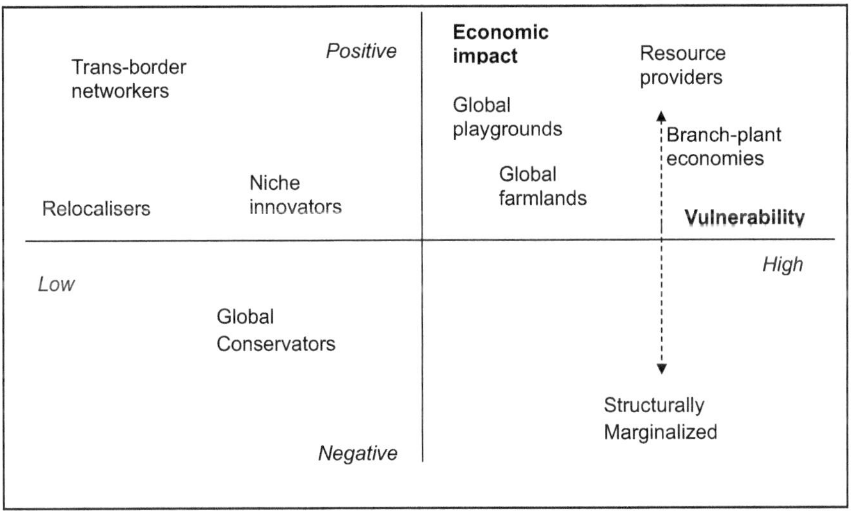


**Figure 12.1 Typology of regional responses to globalization**

are highly vulnerable to external corporate decision-making, and the closure of branch plants can have a severe negative impact on a region's economy. As such, the most effective regional development strategies will support several responses in parallel.

If regional development policies and initiatives are not mobilized to form coherent responses to globalization impacts, regional futures may be left seriously exposed to the vagaries of international events and corporate decisions, and opportunities for economic development might be lost. In some of these cases, a limited capacity to respond to globalization may result from structural disadvantages, such as location, political-economic context and composition of the existing economy. These *structurally marginalized* regions face the greatest challenges from globalization, and risk being drawn into global networks as the 'dumping grounds' for noxious activities unwanted elsewhere (such as some forms of industrial livestock farming, polluting power stations or nuclear waste dumps), or as simply exporters of labour to the global economy. Yet, even in regions facing structural disadvantages – such as Alytus County (see Chapter 7) – there is usually some potential to develop more positive responses to globalization if the appropriate policies and resources can be secured and enacted.

## Strengthening Responses to Globalization

The outcomes of globalization in rural regions are not inevitable. Although both exposure to globalization processes and opportunities to engage with global networks are constrained by regional contexts and characteristics, the interpretative model introduced in Chapter 1 also places an emphasis on the ways in which regions act in response to the globalization impacts they face. Examples of good practice reported in several chapters in this book highlight a number of instances of policies and initiatives that have been proactive in countering the negative impacts of globalization and promoting and taking advantage of positive aspects (see Chapters 6, 7, 9 and 11). Government bodies at the national, regional and local scales have an important role to play in this respect in developing policies and framing and funding programmes for regional development, and we propose some recommendations in respect to these functions later in this chapter. However, the case studies also repeatedly show the critical contribution of individual entrepreneurship and grassroots action in responding to globalization, and hence the significance of civil society as a locus for regional mobilization.

The role of civil society is particularly discussed in Chapter 4 in the context of County Roscommon in Ireland. As the chapter described, civil society groups have been actively incorporated into regional development planning and implementation through partnerships such as LEADER Local Action Groups (also noted to be important in several other regions including Westerkwartier, Comarca de Verín and Alytus county (see Chapters 3, 6 and 7)), as well as in efforts to address social challenges such as the integration of migrants. However, the effective operation of civil society in these activities is conditioned by various factors, including the openness of governmental decision-making, the availability of public funding and other support and interpersonal relations between key actors: this last observation, together with the findings noted earlier about the importance of informal networking between rural businesspeople, in turn points to the importance of social capital as a key component in the capacity of rural regions to respond to globalization, alongside financial capital and environmental capital.

Social capital has been described as 'the glue that holds society together' (Serageldin 1996, p. 196), and has been recognized as a key element in regional and rural development in recent research (Árnason et al. 2009, Dargan and Shucksmith 2008, Parnwell 2007, Serageldin and Grootaert 2000). Mohan and Mohan (2002) argue that social capital can be used to explain economic growth and unequal development as well as the efficiency of ruling institutions, emphasizing how social capital facilitates the construction of networks that permit access to material and non-material resources. As Bohle (2005) puts it, 'within networks of economic actors [...], social capital can enhance the competitiveness of a group with the help of communication and information exchange and this create money capital, real capital and human capital, which could not have been created without affiliation to the network' (p. 69, translation). In this way, social capital is important both to building external, trans-local and trans-national networks, and to strengthening network ties within a region that enhance its capacity to act. This includes by supporting the process of regional learning, as discussed further below.

*Capacity Building, Governance and Knowledge Systems*

The question of how to (best) arrange support for joint learning and innovation in rural grassroots development initiatives, thus empowering them to effectively respond to the challenges and opportunities of globalization, was examined through a final strand of the DERREG research, conducted in six regions (Alytus County, Comarca de Verín, County Roscommon, Oberlausitz, Saarland and Westerkwartier). The details and findings of this research are particularly discussed in Chapters 3 and 6 in relation to the Westerkwartier and the Comarca de Verín.

As described further in Chapter 3, the analysis was underpinned by a conceptual framework that adapted and developed the model of the 'learning region' from its original formulation in urban industrial regions (see Etzkowitz and Leydesdorff 2000, Rutten and Boekema 2007) to fit the setting of rural regions. Tovey (2008) has argued that learning and innovation processes in rural regions require not only support and facilitation for the transfer of technological, expert knowledge and related human capital from knowledge institutions, but also support and facilitation for the use and acquisition of indigenous knowledge about local places and locally embedded resources. Accordingly, the 'learning region' model was modified with 'industry' replaced by 'region' – defined as an arena encompassing diverse actors and commonly represented in learning and innovation interfaces by grassroots development movements; 'the state' replaced by 'public administration' – including all levels from the European Union to individual municipalities; and 'universities' replaced by 'knowledge infrastructure', including universities and colleges, schools, research institutes, state agencies, private consultancies, NGOs and grassroots initiatives themselves (see also Wellbrock et al. 2012, Wellbrock et al. 2013b).

The modified model provided a framework for mapping grassroots learning and innovation initiatives in the study regions, and assisted in identifying factors that are important to the effective functioning of the interfaces between the different components. Once again, this exercise reveals notable variations between the regions (see Burdack et al. 2013, Wellbrock et al. 2013a, Wellbrock et al. 2013b).

For instance, the composition of operational interfaces between organizations and agents involved in supporting and facilitating regional learning and innovation varies in reflection of differences in regional governance structure. These operational interfaces can either be determined by governmental organizations, or designed collaboratively by public and private agents working in partnership. Interfaces can be designed as free-standing associations, public-private partnerships or loose networks, with participants potentially including government agencies, local municipalities, NGOs and local associations, public research or educational institutions, professional service companies, local businesses and individual community members. The shape and scale of an operational interface can affect its visibility within the region and its accessibility to local communities. Similarly, the types of beneficiaries targeted vary between rural regions, reflecting different development priorities in different regional contexts, influenced by demography, geographical characteristics, history and local sense of place; whilst the balance of different types of activities offered will vary depending on the organizational composition of the interfaces and the priorities set by the lead stakeholders.

These differences imply that one cannot simply transfer arrangements and well-working interfaces from one country to another, not even within countries: arrangements have to be contextualised. However, comparative analysis reveals a repertoire of promising practices that can be draw upon as showcases and learned from in improving arrangements and the working of interfaces.

Firstly, the early engagement of local people in reflexive learning exercises can contribute to local ownership over later programmes by encouraging local people to debate and decide their local priorities and hence to inform the focus of subsequent initiatives. The *Brug Toekomst* ('Bridge to the Future') project in the Westerkwatier served this purpose by using students from partner universities to facilitate workshops that brought local organizations together to formulate regional development priorities and objectives. One outcome of the project was the establishment of the 'Westerkwartier Initiative Group', which has become the main operational interface for joint learning, innovation and rural development initiatives in the region (see Chapter 3).

Secondly, stakeholders with leadership and networking roles are important to connect the different 'worlds' of rural development initiatives, public administration and the knowledge support structure. In particular, the involvement of charismatic, informal agents, able to bring together residents with various, possibly conflicting interests, is important. In this way, charismatic, informal agents act as spiders in the networking web. They are generally seen as crucial in forging consensus and in the formation of public-private partnerships that can provide formal working interfaces.

Thirdly, the delegation of tasks and roles within the operational space of an interface is also a crucial factor for successful learning and innovation initiatives, determining the ability of an interface to mediate effectively between the 'worlds' of supporting policies, the knowledge support structure and grassroots development initiatives. To be effective, operators need some delegated decision power in order to attend to the needs of grassroots development initiatives. If not, support may become too bureaucratic or even 'politicalized', demotivating volunteer participants. Grassroots development initiatives need to be able to adapt their development aims to changing political contexts.

Finally, the policies and programmes of public agencies are important in initiating and supporting interfaces for regional learning and innovation, and especially for providing funding. In several of the regions studied, the LEADER programme in particular had played a central role in underpinning interfaces. However, if public administration – at whatever scale – exerts strong control over operational agents and agencies, support and facilitation can turn into clientelism. Thus, the role of public policies as a driving force of place-based development should not be overestimated. The involvement of charismatic, informal agents, usually from within regional civil society, is at least as important in mobilizing participation, establishing formal organizations and developing and sustaining agendas.

### *Strengthening Policies for Rural and Regional Development*

Notwithstanding the above observations about the role of public policies in rural development, there are a number of ways in which the policies adopted and implemented by local, regional and national governments, as well as by the European Commission, could be strengthened to better equip rural regions with the capacity to respond positively and proactively to globalization. Most important of these is the need for policy to be sensitive to geography. A recurrent feature of the analysis presented in this volume has been the variation in experiences and outcomes of globalization in the regions examined, thus policy needs to recognize regional contexts and acknowledge that globalization is not imposed from the top-down, but rather reproduced in place from the bottom up (see Werlen 2007, Woods 2007, 2013).

Institutional capacity is also important for enabling effective policy interventions. Responding appropriately to the specific globalization pressures on particular regions or localities requires an ability to formulate and implement a strategy at the right territorial scale. As such, the presence of empowered regional governance institutions operating across territories that make sense functionally allows for a more targeted and tailored response to pressures from globalization than in states where political authority is either heavily centralized or diffused among small-scale municipalities. Regional capacity to act also tends to be strengthened by the existence of a buoyant regional civil society, including business associations and support networks that can be enrolled in regional development initiatives. Indeed, engaging public authorities, civil society groups and businesses in collective action, for example as part of LEADER partnerships or in regional learning projects, is fundamental to constructing sustainable and inclusive, locally led responses to globalization.

However, some barriers to effectively engaging with globalization opportunities identified in the DERREG research require policy interventions by national or higher-tier regional governments. For example, access to finance capital is a major obstacle for many SMEs wanting to explore international opportunities, especially since the financial crisis of 2008, but requires action by EU and national authorities working with the banks. Similarly, tax concessions and other fiscal incentives that can be offered to attract foreign direct investment to particular regions are commonly the responsibility of national governments.

European and national government policies also need to support transnational networking as a means to open up new international markets for rural businesses, and to capitalize on the opportunities created by the dual affiliations of international migrants. At the same time, such networks need to be firmly embedded in regional contexts to maximize the benefit to regional economies and provide stability. As such, particular importance should be assigned to supporting the expansion of endogenous businesses into international niche markets rather than to attracting foreign direct investment.

In addition to support for economic enterprises and institutions, more attention is required for the integration of international migrants in rural areas. This is of particular importance because migrants no longer move to urban or even metropolitan regions exclusively – which have most often been the focus of attention – but also to rural space, and thus take globalization into the daily experience of its residents. Rural regions can benefit from migration culturally, socially and economically, but policies are required that promote the integration of migrants and counteract discrimination. These initiatives can build bridges between the local residents and migrants, as well as among migrants, and thus contribute to networking and the creation of social capital.

Policies also need to be sensitive to regional contexts and interests in responding to global environmental issues and formulating strategies for sustainable development in rural regions. Undertaken correctly, the valorization of environmental capital has tremendous potential to stimulate and support growth in rural regions as part of an 'eco-economy' that is both economically and environmentally sustainable. The early engagement of local stakeholders is essential to achieving consensus, as is a focus on using endogenous resources that will reinforce regional cultural identity and local supply networks as well as identifying opportunities to connect to niche markets through international networks.

## Conclusion

The research presented in this book describes how the restructuring of rural regional economies in the context of globalization is shaped by the interaction of external processes and actors, and regional capacities, resources and agents. As such, regional development actors can have an influence in shaping the development trajectory of their regions, functioning as catalysts to attract investment, migrants or tourists, or to enable international networking; or formulating and implementing responses to negative globalization impacts. In each case, regional development actors are, in practice, negotiating globalization and crafting strategies that involve the combination of endogenous and exogenous resources, capital and opportunities.

The evidence from the case studies discussed in the preceding chapters not only illuminates the dynamics of globalization in rural regions, and especially the uneven geography of the global countryside, but also provide insights into mechanisms that support effective regional development interventions. As such, four key conclusions can be emphasized.

Firstly, *resilient responses to globalization require support for endogenous business*. Branch plant investments and the exploitation of natural resources to feed the global resource boom can provide a significant injection into regional economies; however, they do not necessarily result in economic growth throughout a region as a whole, and leave a region exposed to the effects of distant political and economic events. More resilient – if not as immediately lucrative – are strategies based on supporting endogenous rural enterprises in taking advantage of globalization to develop international business networks, especially trans-local linkages with equivalent firms in other regions. Whilst the DERREG research records obstacles that smaller rural firms face in engaging international networks, it also highlights the importance of network brokers in facilitating such linkages. The activity of network brokers in connecting rural businesses to international networks – including access to new export markets – can be a catalyst for strategies of 'niche innovation' and 'trans-border networking', whilst network brokering that aims to consolidate linkages within regions can contribute to re-localization strategies.

Secondly, *international migrants can make a significant contribution to rural regional development.* The DERREG research has highlighted the extent and dynamism of international migration to, from and through rural regions in Europe. As such, human capital should not be regarded as a static endogenous resource that needs to be trained up to meet the challenge of globalization, but rather as a fluid resource that needs to be attracted, captured, retained and nurtured. International migrants, including migrant workers, amenity migrants, asylum seekers and return migrants, all possess skills, expertise and experience that can benefit rural economies. In particular, our research points to the potential for entrepreneurship among migrants to contribute to economic development. Yet, with some notable examples, this resource appears to be underutilized in the regional development strategies of rural regions. Out-migrants can also contribute to regional development by acting as bridging agents in constructing international networks. Recognizing the entrepreneurial potential of international migrants can add value to 'global resource providers' and 'global playgrounds' – both of which are associated with in-migration.

Thirdly, *global environmental awareness has created opportunities for the sustainable development of rural environmental capital.* Among the most significant endogenous resources available to rural regions are those associated with the natural environment. Traditionally, rural economies have been built on the extraction of value from natural resources – through mining, industrial forestry and productivist agriculture – but with an environmental cost. Growing awareness of global environmental issues, including climate change and the loss of biodiversity, has increasingly led to these established rural activities being questioned and challenged. The incorporation of global environmental concerns into regional development planning has thus in some cases encountered conflicts and opposition from some elements in rural communities. However, global environmental awareness has also added value to the sustainable use of natural resources, thus creating opportunities for income generation through activities such as renewable energy production, eco-tourism and agri-tourism, the premium branding of sustainable agricultural products and payments for environmental services. These activities underpin the model of the 'global conservators', and can contribute to strategies of 'niche innovation' and 'relocalization', as well offering an alternative way of 'global resource provision' based on renewable energy.

Fourthly, *effective regional development is supported by joint regional learning and innovation.* The inclusive engagement of rural residents and other stakeholders in planning responses to globalization pressures is critical to effective regional development. This can involve reflection on the identity, character and future of a region, in terms of thinking about what needs protecting and preserving, and what endogenous resources might be valorized as part of 'niche innovation' or 're-localization' strategies. Harnessing such reflection as a tool for regional development can be described as a process of regional learning, through which communities become better equipped to respond to challenges and recognize opportunities. In addition to greater awareness, regional learning can also develop technical capabilities, including the rediscovery of lost local knowledge and craft skills. Building capacity for regional learning requires the interface of public administration, knowledge actors and regional civil society, the efficacy of which can itself be dependent on organizational structure, funding arrangements, the distribution of responsibilities and agency and the input of charismatic individual actors.

Europe's rural regions have not been bypassed by globalization, but neither are they helpless victims of globalization processes. Rather, the DERREG research reported in this volume has shown that rural regions across Europe are proactively engaging with globalization processes and their impacts, exploiting new opportunities for international networking and trade and developing responses to the challenges presented by globalization. Rural regions can be independent and innovative actors in global social and economic networks, contrary to some perspectives in EU regional policy that position rural areas as dependent adjuncts to urban economies.

Regional actors are hence participants in the reproduction of global networks through rural regions, and regional development policies and grassroots initiatives consequently can shape regional responses to globalization. For instance, such responses can determine whether regional economies are future-proofed against the sudden withdrawal of foreign branch plants; whether the skills and expertise of international migrants are tapped for economic development; and whether international environmental designations are welcomed as opportunities for sustainable development, or resisted as threats to traditional economic interests.

Developing effective responses to globalization involves reflection and learning, with the successful harnessing of lay and expert knowledge, and the facilitation of entrepreneurship and grassroots action. With appropriate policy support, Europe's rural regions can engage globalization from the bottom up, and can have an influence in shaping their own futures.

## References

Aggarwal, R.M. (2006) Globalization, Local Ecosystems, and the Rural Poor, *World Development*, 34 (8), 1405–18.

Árnason, A., Shucksmith, M. and Vergunst, J. (eds) (2009) *Comparing Rural Development: Continuity and Change in the Countryside of Western Europe*. Aldershot: Ashgate.

Bathelt, H., Malmberg, A. and Maskell, P. (2004) Clusters and knowledge: local buzz, global pipelines and the process of knowledge creation, *Progress in Human Geography*, 28, 31–56.

Bohle, H.-G. (2005) Soziales oder unsoziales Kapital? Das Sozialkapital-Konzept in der Geographischen Verwundbarkeitsforschung, *Geographische Zeitschrift*, 93, 65–81.

Burdack, J., Kriszan, M. and Nadler, R. (2013) Networks of regional learning in rural Eastern Saxony, in Tamásy, C. and Revilla Diez, J. (eds) *Regional Resilience, Economy and Society*, pp. 213–30. Aldershot: Ashgate.

Copus, A., Dubois, A. and Hedström, M. (2011) Expanding horizons: local embeddedness and global engagement among small firms in the European countryside, *European Countryside*, 3, 164–82.

Dargan, L. and Shucksmith, M. (2008) LEADER and innovation, *Sociologia Ruralis*, 48, 274–91.

Dubois, A., Copus, A. and Hedström, M. (2012) Local embeddedness and global links in rural areas: Euclidean and relational space in business networks, in Hedberg, C. and do Carmo, R.M. (eds) *Translocal Ruralism*, 103–21, Dordrecht: Springer.

Düvell, F. (2006) Europäische und internationale Migration: Einführung in historische, soziologische und politische Analysen, in Liebert, U. and Falke, J. (eds) *Europäisierung Beiträge zur transnationalen Europadebatte*. Hamburg: LIT Verlag.

Etzkowitz, H. and Leydesdorff, L. (2000) The dynamics of innovation: from national systems and 'mode 2' to a triple helix of university-industry-government relations, *Research Policy*, 29, 109–23.

Farrell, M., Mahon, M. and McDonagh, J. (2012) The rural as a return migration destination, *European Countryside*, 4, 31–44.

Frys, W. and Nienaber, B. (2011) Protected areas and regional development: conflicts and opportunities – presented on the example of the UNESCO biosphere reserve Bliesgau, *European Countryside*, 3, 208–26.

Frys, W. and Nienaber, B. (2011) Die Situation der Wohnmigranten im ländlichen Saarland, in Hege, H.-P., Knapstein, Y., Meng, R., Ruppenthal, K., Schmitz-Veltin, A. and Zakrzewski, P.(eds) *Schneller, öfter, weiter? Perspektiven der Raumentwicklung in der Mobilitätsgesellschaft*, pp. 94–104, Hannover: Akademie für Raumforschung und Landesplanung.

Granovetter, M. (1985) Economic action and social structure: the problem of embeddedness, *American Journal of Sociology*, 91, 481–510.

Kelly, P.F. (1999) The geographies and politics of globalization, *Progress in Human Geography*, 23, 379–400.

Kitchen, L. and Marsden, T. (2009) Creating Sustainable Rural Development through Stimulating the Eco-economy: Beyond the Eco-economic Paradox? *Sociologia Ruralis*, 49, 273–93.

Kriszan, M. and Burdack, J. (eds) (2013) *Use or protect – environment capital and regional development in the rural areas of Europe: research findings and experiences from the EU project DERREG*. Leipzig: Leibniz-Institut für Länderkunde.

Lampič, B. and Mrak, I. (2012) Globalization and foreign amenity migrants: the case of foreign home owners in the Pomurska Region of Slovenia, *European Countryside*, 4, 45–56.

McCarthy, J. (2008) Rural geography: globalizing the countryside, *Progress in Human Geography*, 32, 129–37.

Mohan, G. and Mohan, J. (2002) Placing social capital, *Progress in Human Geography*, 26, 191–210.

Nadler, R. (2012) Should I stay or should I go? International migrants in the rural town of Zittau (Saxony) and their potential impact on rural development, *European Countryside*, 4, 57–72.

Nadler, R., Kriszan, M., Nienaber, B. and Frys, W. (2012) Zuwanderung internationaler Migranten in schrumpfende ländliche Regionen: die Fallbeispiele Ostsachsen und Saarland. *Europa regional*, 18.2010 (2012) 2/3, 107–21.

Nienaber, B. and Frys, W. (2012) International labour migration in European rural regions: the example of Saarland, Germany, *European Countryside*, 4, 73–88.

Nienaber, B. and Kriszan, A. (2013) Entgrenzte Lebenswelten: Wohn- und Arbeitsmigration als Ausdruck transnationaler Lebensentwürfe im deutsch-luxemburgischen und deutsch-polnischen Grenzraum. *Raumforschung und Raumordnung*, 71(3), 221–32, DOI: 10.1007/s13147–013–0230–2.

Ni Laoire, C. (2007) The 'green grass of home'? Return migration to rural Ireland, *Journal of Rural Studies*, 23, 332–44.

Parnwell, M.J.G. (2007) Neolocalism and renascent social capital in northeast Thailand, *Environment and Planning D: Society and Space*, 25, 990–1014.

Potočnik Slavič, I. (2011) The relevance of linking favourable business environments with embeddeness: SMEs in rural areas of the Goriška Region, Slovenia, *European Countryside*, 3, 183–97.

Rutten, R. and Boekema, F. (2007) *The Learning Region: Foundations, State of the Art, Future*. Cheltenham: Edward Elgar.

Serageldin, I. (1996) Sustainability as opportunity and the Problem of Social capital, *The Brown Journal of World Affairs*, 3, 187–203.

Serageldin, I. and Grootaert, C. (2000) Defining social capital: an integrating view, in Dasgupta, P. and Serageldin, I. (eds) *Social Capital: A multifaceted perspective*, pp. 40–58. Washington, DC: World Bank Publications.

Šťastná, M., Kniezková, T. and Náplavová, M. (2011) Embedding of rural businesses in the South Moravian region, *European Countryside*, 3, 198–207.

Steger, M. (2003) *Globalization: A Very Short Introduction*. Oxford: Oxford University Press.

Tödling, F. and Kaufmann, A. (1999) Innovation systems in regions of Europe – a comparative analysis, *European Planning Studies*, 7, 699–717.

Tovey, H. (2008) Introduction: Rural sustainable development in the knowledge society era, *Sociologia Ruralis*, 48, 185–99.

Wellbrock, W., Roep, D. and Wiskerke, J. (2012) An integrated perspective on rural regional learning, *European Countryside*, 4, 1–16.
Wellbrock, W., Nienaber, B., Kriszan, M., Nadler, R., Frys, W. and Roep, D. (2013a) Arranging support for collective learning in rural areas of Germany, in Tamásy, C. and Revilla Diez, J. (eds) *Regional Resilience, Economy and Society*, pp. 195–212. Aldershot: Ashgate.
Wellbrock, W., Roep, D., Mahon, M., Kairytė, E., Nienaber, B., Dominguez Garcia, M.D., Kriszan, M. and Farrell, M. (2013b). Arranging public support to unfold collaborative modes of governance in rural areas, *Journal of Rural Studies*, 32, 420–429.
Werlen, B. (2007) *Globalisierung, Region und Regionalisierung. Sozialgeographie alltäglicher Regionalisierungen.* Stuttgart.
Woods, M. (2007) Engaging the global countryside: globalization, hybridity and the reconstitution of rural place, *Progress in Human Geography*, 31, 485–507.
Woods, M. (2013) Rural development, globalization and European regional policy: perspectives from the DERREG project, *Geographia Polonica*, 86, 99–110.
Woods, M. (2014) Regions engaging globalization: a typology of regional responses in rural Europe, *Journal of Rural and Community Development*, 8, 113–26.

# Index

For Product Safety Concerns and Information please contact our EU
representative GPSR@taylorandfrancis.com
Taylor & Francis Verlag GmbH, Kaufingerstraße 24, 80331 München, Germany

www.ingramcontent.com/pod-product-compliance
Ingram Content Group UK Ltd.
Pitfield, Milton Keynes, MK11 3LW, UK
UKHW022134190926
475702UK00013B/1797
* 9 7 8 1 1 3 8 5 4 6 5 1 6 *